AN INTRODUCTION
TO INVERTEBRATE
ENDOCRINOLOGY

AN INTRODUCTION TO INVERTEBRATE ENDOCRINOLOGY

A. S. TOMBES

Division of Biological Sciences
Clemson University
Clemson, South Carolina

 1970

ACADEMIC PRESS   NEW YORK and LONDON

115053

592
$T656$

Academic Press, Inc.
111 Fifth Avenue, New York, New York 10003

United Kingdom Edition published by
Academic Press, Inc. (London) Ltd.
Berkeley Square House, London W1X 6BA

Library of Congress Catalog Card Number: 72-91428

Printed in the United States of America

To the team at 257

CONTENTS

PREFACE

This introductory work presents invertebrate endocrinology as a subject in itself rather than as a subdivision of vertebrate endocrinology or physiology. It is my belief that the subject has reached a point in its growth, measured either in the absolute amount of information available or in the apparent interest from biologists, that a separate publication can stand alone. Such a weaning process should prove to be a stimulus for further instruction and investigation into the endocrine bases underlying the biology of invertebrates. Arthropods have received the major share of the attention given to invertebrates, and several monographs dealing with insects and crustaceans are available. A number of endocrinology and physiology textbooks as well as advanced treatises and reviews on neurosecretion or neuroendocrinology have dealt at some length with invertebrates. Thus the student of endocrinology and physiology has been favored with the availability of information while the student of invertebrate zoology, unless he has searched for the information himself or has been directed to the advanced texts, has been somewhat slighted.

Thus this book, written principally for the student zoologist, is so arranged as to follow the zoologist's approach to a subject, phylum by phylum, rather than that of the physiologist, system by system. By progressing phylogenetically we can observe one very pronounced common denominator throughout all the phyla: the neurosecretory cell, which appears histologically and functionally quite similar, from hydra to insects. Martin Wells, in his book "Lower Animals," cautions one not to expect similarity of function among phyla when studying a physiological system, but to be mindful of the significance when similarity does appear. "The occurrence of a similar mechanism in two unrelated groups may be a matter of chance. If the same sort of machinery turns up in several, one is entitled to interpret the recurrence as an indication that there is, perhaps, only one means of solving this particular problem with biological material. Nerve and muscle are two such cases." With this observation in mind, we emphasize that only after further study of the invertebrates can we assess more completely the phylogenetic significance of the neurosecretory cell.

Of the two major divisions in the book, Part 1, comprised of Chapters 1 and 2, covers the general mechanisms and structure of the endocrine system. The second part, consisting of Chapters 3 through 11, is a discussion of the endocrine system of seven phyla. The first four—Cnidaria, Platyhelminthes, Nemertinea, and Nematoda—can be considered as the minor group with the last three phyla—Mollusca, Annelida, and Arthropoda—constituting the major group. Since the neurosecretory cell is the dominant endocrine component of the invertebrates, a limited morphological discussion of the nervous system for each phylum is necessary. Thus morphology receives initial attention in each chapter, including, when available, ultrastructural evidence to augment the existing information. A discussion of the physiological processes, i.e., reproduction, growth and development, osmoregulation, and others which have been recorded as being under neuroendocrine influence, follows the morphology presentation. It is only after the neuroendocrine elements have been localized and described histologically that a control mechanism for a physiological process can be hypothesized and tested experimentally.

The aim of this book, therefore, is to discuss in general terms the structure and function of invertebrate endocrine systems as they represent integrating mechanisms between stimuli of extrinsic and intrinsic origins and physiological response systems of the organism. It is my hope that this presentation will make young biologists aware of the wealth of problems among the invertebrates which can be tackled by employing the descriptive and experimental methods now available.

Bern and Hagadorn's chapter in Bullock and Horridge's "The Structure and Function of the Nervous System of Invertebrates" (1965), the English edition of "Neurosecretion" by Gabe (1966), "L'endocrinologie des Vers et des Mollusques" (1967) by Durchon, "L'endocrinologie des Insectas" (1968) by Joly, "Invertebrate Zoology" by Meglitsch (1967), and "Bibliographia Neuroendocrinologica" edited by M. Weitzman have proved invaluable in the preparation of much of this material.

A note of appreciation is due the staff of the Clemson University library, especially Miss Peggy Hopkins and the late Mr. John Goodman, for their kind and capable help in procuring material which was not in their collection. I am also grateful to Miss Cynthia Warner for filing and unfiling many items of information, to Drs. George Folkerts and Donald Forrester who read portions of the manuscript and provided many helpful suggestions, and to my wife for masterfully typing the original manuscript. My interest in this topic was greatly expanded by my association for a year, 1965–1966, as a postdoctoral fellow with Drs. Dietrich Bodenstein and David Smith; to them my sincere appreciation is extended as well as to the National Institutes of Health, the provider for that year of study. I would also like to express my sincere thanks to the staff of Academic Press for their aid in producing this work.

For errors that remain or any uneven emphasis on certain taxonomic groups and topics, I am responsible and will welcome any comments by the interested reader.

PART 1

PRINCIPLES OF ENDOCRINOLOGY

CHAPTER 1

HORMONAL MECHANISMS
OF INTEGRATION

One of the more exciting aspects of modern biology is the knowledge of how individual cells are regulated. The sequences involved in the control of protein synthesis, for example, are now commonplace in the curriculum for beginning biologists. Of no less importance is the regulation of each organ system and the complete coordination of all physiological processes. Although the ideas did not all originate with W. B. Cannon, he, nevertheless, synthesized the concept of homeostasis which contends that an organism maintains a certain constancy in all systems of the body, thus taking on the characteristics of a steady state. This ability to maintain constancy is observed to a high degree of development in the mammals through the integration of nervous and hormonal systems, and to a lesser degree in the lower vertebrates. The invertebrates also have similar mechanisms of regulation, although they are considerably less exacting. Thus it appears characteristic of all physiological processes to operate within certain limits which become more narrow, and the regulating mechanisms more exact, the higher the organism appears on the phylogenetic tree.

The endocrine and nervous systems are integrative in nature, for together they provide the organism with sophisticated external and internal sensing devices which feed information into the central nervous system for analysis and integration, and in turn orchestrate the necessary target systems to conduct the appropriate responses efficiently. The nervous system is characterized by its ability to respond

3

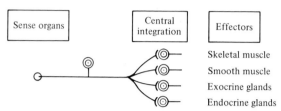

Fig. 1-1. Diagram of the reflex arc, showing the effectors by which the integrative action of the nervous system is brought about. (From Ganong, 1966.)

to stimuli with a high-speed and short-duration response. The opposite is more characteristic of the endocrine system, for a certain concentration of an active chemical must be obtained in the blood and maintained for a period of time before the response is detected. Thus the endocrine system, when compared to the nervous system, is involved in responses that are slower and longer lasting. Since the nervous system is the sole sensing arm of this integrating reflex, the endocrine organs then can be referred to as effector units of the nervous system; the link between the two is therefore the seat of integration. In Fig. 1-1 this link is referred to as central integration, and in either vertebrates or invertebrates it consists of the doubly specialized neurosecretory cell (Bern, 1963).

Through their neuroendocrine systems, all animals from hydra to man are not only able to regulate growth and development leading to satisfactory reproduction but are able to detect changes in their external environment and bring about appropriate adjustment in the body chemistry. Thus the preservation of a species under stress—the maintenance of homeostasis—might be a reflection of the efficiency by which the neuroendocrine system in that organism is able to detect and direct the appropriate corrective action.

I. NEUROSECRETION

In recent years the traditional distinctions between endocrine and nervous systems have diminished, and the concept of the two systems working together to provide the organism with an integrative network has emerged. This has been fostered by the observations that nerve cells synthesize, transport, and secrete chemicals, which is broadly the definition of neurosecretion. The neuroanatomist has been aware of this for some time, for the synapse refers to the special locus of contact between two neurons where excitatory or inhibitory influences are transmitted from one neuron to another (De Robertis, 1964). This particular type of neurosecretion is specifically for the production of neurohumoral materials or neurotransmitters, i.e., acetylcholine, adrenaline, and noradrenaline, and is an essential concomitant of the impulse transmission which is of millisecond duration and travels across a distance of several millimicrons (Table 1.1). Since our concern is not with this aspect of neurosecretion, it will not be discussed further.

From the classic investigation of Bragmann, Hanstrom, and the Scharrers has come the second and more commonly used concept of neurosecretion. This concept refers to a nerve cell containing prominent, stainable inclusions which appear to represent hormonally active secretory material, synthesized within the cell body, often transmitted via an axon to be released from the cell at some distance from the site of synthesis, and remaining active for an extended period of time (Fig. 1-2). This theory of neurosecretion has recently been subjected to serious evaluation as more findings from various representatives of the animal kingdom are made known, but the basic concepts still hinge on the secretory activity of nerve cells which can be related to measurable physiological effects. Histological observations, which have relied heavily upon Gomori's classic staining methods or their modifications, are increasingly being questioned as adequate means for evaluating neurosecretion (Bern, 1966; Bern and Knowles 1967; Knowles and Bern, 1966). A number of tissues or cellular organelles completely foreign to neurosecretion have responded positively to the selected stains, and the possibility always exists of a significant secretion for which no distinctive staining reaction is as yet available. This was shown recently to be the case in an insect where the neurosecretion is not reactive to paraldehyde-fuchsin on chrome-hematoxylin-phloxine, but stains intensely with azan (Raabe, 1965).

The electron microscope has provided an ultrastructural understanding upon which to base the interpretation of this stainable secretory material, and is now being used as an additional qualitative test for neurosecretion. The topic of ultrastructure will be discussed at greater length in the next chapter, but it may be mentioned here that the basic neurosecretory product, the elementary neurosecretory granules, appear to vary in electron density, are membrane-bounded, average between 1000 and 3000 Å in diameter, and have their origin in the Golgi apparatus of the neurosecretory cell.

TABLE 1-1

Classification of Invertebrate Hormones and Hormone-like Secretions[a]

Hormone type	Site of elaboration	Mode of action	Example
Neurohumors	Nerve cells	Mediators of nerve transmission; travels a very short distance and acts during a brief period	Acetylcholine 5-Hydroxytryptamine Noradrenaline Adrenaline
Neurohormones	Nerve cells	Coordinator of neuroendocrine function; may be transported long distance and stored in neurohemal organs	Arthropod Brain hormone
Glandular hormones	Tissues other than the nervous system	Classic endocrine action; may be transported throughout body to target organs	Ecdysone Juvenile hormone
Pheromones	Glands opening to the outside	Providing social organization at extremely low concentrations following release to the external medium	Bombykol Gyptol

[a] Tabulated from Gabe et al., 1964.

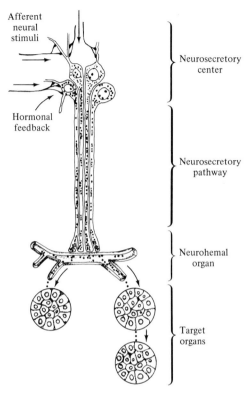

Afferent
neural
stimuli

Hormonal
feedback

Neurosecretory
center

Neurosecretory
pathway

Neurohemal
organ

Target
organs

Fig. 1-2. Generalized diagram of a neurosecretory system. The active principle originates in the perikarya, is transported to the nerve terminals by axoplasmic flow, and is there released into the general circulation for the stimulation of target organs. The neurosecretory cells receive nervous impulses from external as well as internal sources, which includes hormonal feedback. (After Scharrer and Scharrer, 1963.)

Electron microscopy also has its sources of error, since a number of cells contain components such as phospholipoprotein droplets of lysosomelike bodies which are often indistinguishable from the elementary neurosecretory granules. Synaptic vesiclelike inclusions in neurosecretory axons have provided problems over which considerable debate has raged. Nevertheless, in the face of these potential pitfalls, we have yet to find better histological diagnostic tools than the improved stains of Gomori and the finding, by electron microscopy, of elementary neurosecretory granules.

Our total understanding of neurosecretion is still meager, but, with time, techniques certainly will be made available to categorize the neurosecretory cells and their products into types, associating specific functions with each. A number of attempts have been made in this direction. Knowles (1965) suggested two categories of neurosecretion, common in both vertebrates and invertebrates, which appear at the ultrastructural level: type A, with granules greater than 1000 Å in diameter, and peptide in nature, and type B, with granules less than 1000 Å in diameter, and with a nonpeptide or possibly an amine secretion. Type A is represented by the more commonly reported category of neurosecretion. On the level of light microscopy there are almost as many schemes for cell classification as there have been histological studies; these schemes will be mentioned where appropriate for each phylum throughout the book.

Thus the note of caution which will be repeated often is that stainable inclusions at the light microscope level and elementary neurosecretory granules at the electron microscope level are certainly strong diagnostic features but should not be the sole points on which a potential neurosecretory system is judged. If physiological activities are to be ascribed to a system, then variations in cytology reflecting some secretory cycle should be detectable.

In this connection the following six conditions of proof, widely followed in vertebrate endocrinology for the establishment of legitimacy of an endocrine organ, are listed here since they are also appropriate for the invertebrate system: (1) Identify a tissue or organ in which a hormone is being secreted. (2) Demonstrate that a correlation exists in the variation in appearance of the secretory cells with a demand made for the hormone in the organism. (3) Attempt to remove or immobilize the tissue with the appearance in the body of clearly defined signs of hormone deprivation. (4) Remove this deficiency by replacement therapy of tissue implants or tissue extracts. (5) Demonstrate the presence of the hormone in the circulating medium. (6) Purify, characterize, and synthesize the hormone so that further analysis of the hormonal-target organ relationship can be studied.

II. NEUROENDOCRINE INTEGRATION

The ability of an organism to react to environmental signals, to respond to those which are important, to disregard the irrelevant ones, and to retain a condition of normalcy throughout the body is a credit to the integrative action of the central nervous system. This essentially involves three interrelated units: afferent pathways, integrative centers, and efferent pathways (Scharrer, 1966).

A. Afferent Pathways

The prime movers in any reflex arc are the receptor cells which send their signals by way of the afferent axon into a ganglion for interpretative action. This input can originate in organs of special senses, i.e., photoreceptors, chemoreceptors, and acoustic apparatus, or from touch, pain, and temperature receptors. Stimuli may also result from changes of the internal environment, i.e., pH, osmolarity, chemical composition of blood, or, indeed, from hormonal feedback. A considerable volume of information has been obtained and catalogued concerning the area of input by the sensory physiologists and is readily available in appropriate textbooks. Two examples are given here which illustrate this aspect of integration. First, initiation of the molting cycle in nymphs of the blood-sucking hemipteran, *Rhodnius*, depends upon nervous stimulation having its origin in stretch receptors in the abdomen (Fig. 1-3). Following a sufficiently large blood meal the receptors are activated, and through the ventral nerve cord stimulate the neurosecretory cells of the brain to produce their appropriate secretions. Second, the octopus exercises hormonal control of gonad maturation by sensing the length of photoperiod and passing a stimulus to the brain via the optic nerve (Fig. 7-13).

B. Integrative Centers

Nervous and chemical afferent stimuli, by way of neural and vascular routes, converge on the central nervous system which, in turn, transmits excitatory or inhibitory signals to the target organs. That portion of the nervous system which receives bioelectric impulses and translates them into effective nervous or hormonal output signals has been termed the "final common pathway" (Scharrer, 1965). In context of invertebrate endocrinology, this takes the form of the perikaryon of neurosecretory cells, the link in the chain uniting nervous and endocrine systems. Cells, in response to the received stimuli, release one or more neurohormones which stimulate or inhibit certain metabolic processes of cells in other tissues. In the aforementioned case of *Rhodnius*, the integrative center receives the stimuli

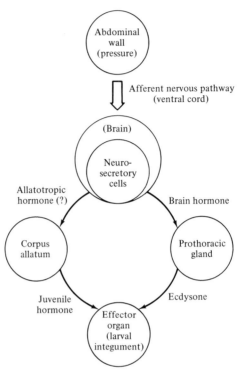

Fig. 1-3. Illustration of the neuroendocrine control of molting in insects. Afferent stimuli from the abdominal wall elicit the release of neurosecretions which stimulate the prothoracic gland to release ecdysone and the corpus allatum to release the juvenile hormone. (After Scharrer and Scharrer, 1963.)

from stretch receptors and, in turn, synthesizes and releases the correct neurohormones in the proper sequence and at the proper concentration required to initiate the process of molting.

C. Efferent Pathways

Neurohormones may act either directly on a target organ or on an endocrine tissue which in turn influences the target organ. Six pathways are illustrated in Fig. 1-4 which show possible relationships between neural and hormonal components (Frye, 1967; Hagadorn, 1967).

1. FIRST-ORDER NEUROENDOCRINE REFLEX

In the lower phyla most neuroendocrine reflexes are first-order, with the neurosecretory cells exerting direct control over target structures (Fig. 1-4, B_2). Two examples would be the inhibition of gonadal

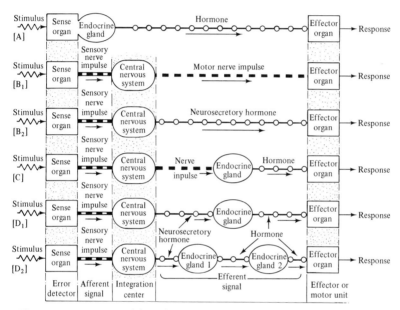

Fig. 1-4. An expansion of the reflex arc showing six possible means of activating a motor unit following the receipt of a stimulus by a sense organ. (A) The endocrine gland is also the sensory unit and responds directly to a stimulus (change in internal chemistry) by secreting a hormone, which in turn elicits a response by the effector organ. (B) Sensory information is transmitted to the central nervous system where it is integrated with other sensory input into the central nervous system, which then directs an appropriate efferent signal. B_1 illustrates a typical neural reflex arc in which the efferent signal is a motor nerve impulse. In B_2 the efferent signal is a hormone produced by neurosecretory cells in the central nervous system. The position of the neurosecretory cells is analogous to that of the motor nerve in the overall reflex. (C) The efferent signal is propagated in two stages: first a motor impulse emanates from the central nervous system which then stimulates an endocrine gland to secrete a hormone that elicits the final response. D_1 is analogous to C, but both stages of the efferent pathway are hormonal. Neurosecretory cells secrete a hormone that activates an endocrine gland. D_2 represents a further level of complexity in which, beginning with the neurosecretory hormone, three different hormones are successively involved in the efferent pathway. B_2, D_1, and D_2 illustrate different types of neurosecretory responses referred to in the text as first-, second-, and third-order neuroendocrine reflexes. (After Frye, 1967.)

maturation in annelids by direct release into the hemolymph of a neurohormone (page 82), and the control of the light-adaptive distal retinal pigments in the eyes of crustaceans (page 136). The possibility also exists that the stimuli may act directly on neurosecretory cells within the central nervous system, which would make the B_2 scheme of Fig. 1-4 more like scheme A.

2. SECOND-ORDER NEUROENDOCRINE REFLEX

The incorporation of one nonneuronal endocrine organ between the neurosecretory cells of the central nervous system and the final target organ illustrates this second-order reflex (Fig. 1-4, D_1). This

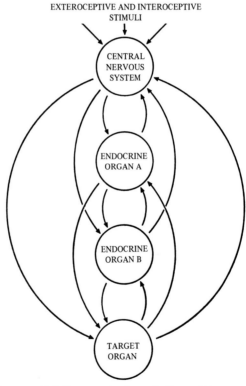

Fig. 1-5. A diagram which illustrates the theoretically possible neuroendocrine interactions to include the first-, second-, and third-order systems. The arrows indicate both nervous and hormonal connections. No species utilizes all available connections. (From Scharrer and Scharrer, 1963.)

additional structure is most often in close association with the neuro-secretory cells or processes of such cells. This positioning facilitates efficient transfer of information which might not be the case if the active components entered the vascular system. Again, the *Rhodnius* example of Fig. 1-3 illustrates two second-order reflexes with neuro-hormones influencing the prothoracic glands and corpora allata to subsequently release ecdysone and juvenile hormone, respectively.

3. THIRD-ORDER NEUROENDOCRINE REFLEX

Although a valid example of this more complex reflex has yet to be found in the invertebrates, the possibility is good that it does exist. As more invertebrate endocrine systems are found and their intricate relationships are explored, it is possible that two nonneural endocrine organs might exist between the neural control and the ultimate target organ (Fig. 1-4, D_2).

Figure 1-5 illustrates, by way of summary, the various possibilities for hormonal integration within an organism. The neurosecretory cell of the central nervous system becomes the final common path be-tween a multiplicity of possible afferent stimuli reaching the central nervous system and the physiological activities which are controlled by these stimuli, often by way of one or two endocrine organs. The possibility of feedback mechanisms existing at similar levels is also noted.

References

Bern, H. A. (1963). The secretory neuron as a doubly specialized cell. *In* "General Physiology of Cell Specialization" (D. Mazia and A. Tyler, eds.), pp. 349–366. McGraw-Hill, New York.

Bern, H. A. (1966). On the production of hormones by neurons and the role of neuro-secretion in neuroendocrine mechanisms. *Symp. Soc. Exptl. Biol.* **20,** 325–344.

Bern, H. A., and Knowles, F. G. W. (1967). Neurosecretion. *Neuroendocrinology (N.Y.)* **1,** 139–186.

De Robertis, E. D. P. (1964). "Histophysiology of Synapses and Neurosecretion." Macmillan, New York.

Frye, B. E. (1967). "Hormonal Control in Vertebrates." Macmillan, New York.

Gabe, M., Karlson, P., and Roche, J. (1964). Hormones in invertebrates. *Comp. Bio-chem.* **6,** 246–298.

Ganong, W. F. (1966). Neuroendocrine integrating mechanisms. *Neuroendocrinology (N.Y.)* **1,** 1–13.

Hagadorn, I. R. (1967). Neurosecretory mechanisms. *In* "Invertebrate Nervous Systems" (C. A. G. Wiersma, ed.), pp. 115–124. Univ. of Chicago Press, Chicago, Illinois.

Knowles, F. G. W. (1965). Neuroendocrine correlations at the level of ultrastructure. *Arch. Anat. Microscope. Morphol. Exptl.* **54**, 343–358.

Knowles, F. G. W., and Bern, H. A. (1966). Function of neurosecretion in endocrine regulation. *Nature* **210**, 271–273.

Raabe, M. (1965). Recherches sur la neurosecretion dans la chaine nerveuse ventrale du Phasme *Clitumnus extradentatus:* Les epaississements des nerfs transverses, organes de signification neurohemale. *Compt. Rend.* **261**, 4240–4243.

Scharrer, E. (1965). The final common path in neuroendocrine integration. *Arch. Anat. Microscop. Morphol. Exptl.* **54**, 359–370.

Scharrer, E. (1966). Principles of neuroendocrine integration. In "Endocrines and the Central Nervous System." *Res. Publ. Assoc. Res. Nervous Mental Disease* **43**, 1–35.

Scharrer, E., and Scharrer, B. (1963). "Neuroendocrinology." Columbia Univ. Press, New York.

CHAPTER 2

STRUCTURE OF THE
ENDOCRINE ORGANS

The endocrine organs of invertebrates can be divided into two categories: those with their ontogeny in the nervous system (neurosecretory cells), and those of nonneural origin (ductless ectodermal glands). In this chapter the generalized structure of both types will be discussed, utilizing evidence from light and electron microscopy (Charniaux–Cotton and Kleinholz, 1964; Gabe, 1966; Hagadorn, 1967a; Scharrer, 1965).

I. NEUROSECRETORY CELLS

Since many neuronal features such as axis cylinders, neurofibrillae, Nissl bodies, and synaptic vesicles are present in neurosecretory cells, it has been assumed by many observers that such cells are modified neurons rather than glia cells which have become part of the nervous system. In this case, the ability to transmit electrical impulses is the initial function of these cells; the neurons later acquired the function of integrating afferent stimuli into a chemical molecule which, when released from the axon, is capable of performing as a hormone. In light of the same evidence of similarity between the neurosecretory cell and the typical neuron, and because neurosecretory cells have been observed in very primitive metazoa, others have reasoned that the first nerve cells evolved as neurosecretory transport axons, and not for the conduction of excitation. It is further reasoned that from this primitive function, chemical transmission could have easily evolved (Horridge, 1968). To accept one idea over the other, we must wait for further study and evaluation of the more primitive nervous systems.

Since the early description by Hanstrom (1931) of "neuroglandular" cells in Crustacea, many diverse histological reactions have been employed to demonstrate neurosecretion. Gomori can be credited with two of the more acceptable and most widely used stains: chrome–hematoxylin, and paraldehyde-fuchsin. Both of these are basic stains; the material which possesses an affinity for these stains has been referred to in the literature by the rather imprecise adjective of "Gomori-positive." The acid stains of phloxine, light green, chromotrope 2R, and Orange G are classed as "Gomori-negative." Although paraldehyde-fuchsin is probably the most commonly used stain, the prudent investigator relies upon more than one stain for the identification of suspected neurosecretory cells and attempts to determine some functional significance of these cells and their products by noting secretory cycles. One should also remember that these techniques are not staining the neurohormone but, rather, the carrier substances (lipoprotein, glycoprotein, glycolipoprotein) which are in association with the much smaller hormone molecule.

The requirements in time and equipment need not be extensive to observe neurosecretory cells in the living state, for the cells of many arthropods possess a characteristic bluish-white opacity when viewed *in situ* under saline. Recently a technique has been developed for staining neurosecretion in whole mounts of nervous tissue, thus providing a more rapid means of observing the neuroendocrine system. This procedure has found initial acceptance in the study of insect neurosecretion, but possibilities exist throughout the invertebrate phyla (Dogra and Tandan, 1964).

With the application of electron microscopy to the identification of neurosecretory cells one additional criterion has been added, i.e., that of the "elementary neurosecretory granule" (Fig. 2-1). These granules appear to be membrane-bounded, and a number of questions have arisen as to the origin and destination of the membrane once hormone release from the axon has occurred. The contents of this single vesicle vary from electron-dense to pale, almost as if the membrane's contents had been released but without an appreciable change in diameter. Unfortunately, the term "granule" carries the connotation of an inorganic, inactive particle. This is not the case, for there are

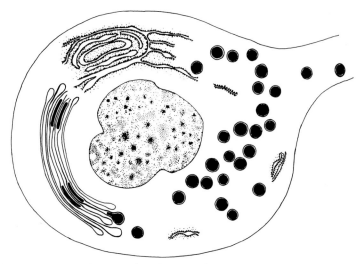

Fig. 2-1. Diagram of presumed perikarya synthesis of neurosecretory material, with endoplasmic reticulum contributing protein "raw material" to the Golgi apparatus for packaging into elementary neurosecretory granules. (From E. Scharrer and Brown, 1962.)

indications that the membrane actively participates in the formation and possible maturation of the granule's content.

A. Synthesis of Neurosecretory Material

Evidence accumulated in recent years through the use of electron microscopy has greatly increased our knowledge concerning this synthetic process. Thus, based on ultrastructural evidence, the following hypothetical scheme is presented. The protein component of the neurosecretory substance has its origin in the rough endoplasmic reticulum. It then moves into the Golgi region and is fabricated into a membrane-limiting unit. In Fig. 2-2, Bern (1963) shows two ways this formation may occur and suggests, with other authors, that the vesicle membrane may not serve simply as an inactive limiting membrane, but may possess dynamic properties aiding still further in the modification of the vesicle content after it has passed from the Golgi. The neurosecretory vesicle is then somewhat endowed with properties of a synthetic organelle, i.e., it is able to change the quali-

Fig. 2-2. Possible modes of formation of elementary neurosecretory granules by the Golgi apparatus, based on ultrastructural studies. Diagram on left represents appearance of electron-dense material within Golgi lamellae and the formation of membrane-limited granules by budding and vesiculation. Diagram on right represents formation of electron-lucent vesicles and their subsequent transformation into electron-dense granules. (From Bern, 1963.)

tative and quantitative nature of its contents as it passes from the cell body to its point of release.

B. Axonal Transport of Neurosecretory Material

Bern and Hagadorn (1965) summarized three principal lines of evidence for the existence of axonal transport of neurosecretory materials: (1) cytological observation of stained material within axons as Herring bodies or in a "string of pearls" fashion, (2) accumulation of secretory material proximal to a transection of an axon, and (3) *in vivo* observation of secretion moving along axons. When this evidence is then coupled with the accepted concept of axoplasmic flow, neurosecretion transport from the site of synthesis to the area of release is an accepted process. During this transport, adjustment in the quality and quantity of material within the vesicle may occur which might be detected by a change in size and electron density of the granule.

Even if transport of the formed granules were deemed unnecessary, or even unlikely in certain axons, it would still be necessary to account for the movement of precursors to allow for local production of neurosecretory granules at the distal regions of the axon. However, the weight of evidence from vertebrate as well as invertebrate material would definitely indicate axoplasmic streaming of the neurosecretion in a distal direction.

C. Release of Neurosecretory Material

After neurosecretory synthesis and transport occur, release from the axon is imperative. This may involve the release of the entire transported material or just the active substance which was in association with the elementary granule. It is believed that this discharge generally takes place at the axon terminals, but considerable evidence indicates that it may occur elsewhere, either in the cell body or along the axon. Considerable uncertainty is encountered in a discussion of the causes of release and the fate of the stainable material following release; however, the three main views on the mechanism of release were summarized by Hagadorn (1967b) and are shown in Fig. 2-3. The most direct route out of the axon would be the passage of the intact granule across the membrane into the circulatory medium. This would require that it pass along the intact membrane, or across the membrane after it has been disrupted in some way. Evidence in support of this idea has been obtained with earthworms and insects.

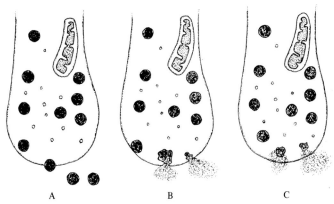

A   B   C

Fig. 2-3. Possible modes of release of neurosecretory material from axon endings. (A) Release of intact granule into the vascular system. (B) Release of the contained material into the vascular system by exocytosis or reverse pinocytosis. (C) Diffusion of the active principle across the membrane of the neurosecretory granule as well as axon membrane. (After Hagadorn, 1967b.)

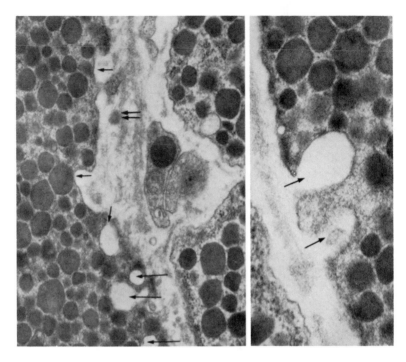

Fig. 2-4. Two views of a possible neurosecretory release mechanism in the corpus cardiacum of *Carausius*. (A) Two axons are separated by an extracellular space with the axon surface at several points (single arrows) indented to form cavities which may represent the site of release. A cohesive secretion droplet may be the dense body indicated by the double arrows. × 38,000. (B) A similar field but at higher magnification illustrating two omega-shaped depressions. × 72,000. (From Smith and Smith, 1966.)

A second release procedure is that of exocytosis or reverse pinocytosis. This occurs when the granule and axon membranes fuse, thereby emptying the hormonally active material into the intercellular space. This has been reported in the corpus cardiacum of several insects (Fig. 2-4).

The third possibility would require that the secretory material diffuse across the axonal membrane after it has made its way out of the granule's limiting membrane. This would partially account for the presence of small, pale, "ghost" granules or synapticlike vesicles, often observed in the vicinity of the supposedly disrupting neurosecretory granule (Fig. 2-5).

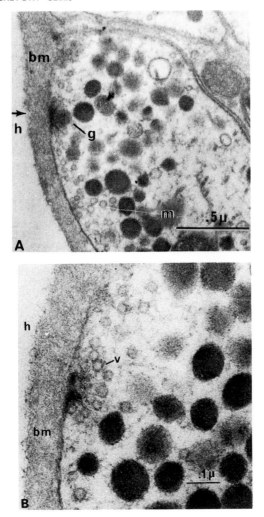

Fig. 2-5. Neurosecretory axcns adjacent to the basement membrane of the corpus cardiacum of *Calliphora*. (A) A disintegrating large granule just beneath the plasma membrane with an accumulation of small vesicles nearby. bm, basement membrane; g, granule; h, hemolymph; m, mitochondria. (B) An adjacent section to A and at higher magnification showing the small synaptic-like vesicles in high concentration at the suspected site of neurosecretion release. v, vesicle. (From Johnson, 1966.)

D. Neurohemal Organs

A decade ago the term "neurohemal organs" was introduced to describe the association of expanded neurosecretory axon terminals with vascular structures. Examples of these organs are the sinus glands and postcommissural organs of crustaceans, the cerebral glands and connective bodies of myriapods, the corpora cardiaca of insects, and the neurohypophysis of vertebrates. Neurohemal organs were then considered solely for storage and release of neurosecretory material. Even though this is an appropriate and quite useful term, there are several implications in this definition which deserve further comment. Many neurohemal organs also may possess intrinsic neurosecretory cells. B. Scharrer (1963) was the first to provide a detailed ultrastructural analysis of an insect's neurohemal organ, the corpus cardiacum, and described cells with endocrine activity similar to the neurosecretory cells in the protocerebrum. Products of the cardiacum's neurosecretory cells may then be released from the surface of the neurohemal organ or extend to other tissues by way of axonal transport.

The other implication is that all neurosecretory axons terminate in a neurohemal structure. Many axons may pass through such structures without any release occurring and then continue on to innervate a distant target organ. In other words, a neurohemal organ, per se, is not a necessary component of a functional neurosecretory system.

Much has been written recently about the limitations which are imposed on any evaluation of a neurosecretory cell when only the concentration or density of stain in histological sections is used as an indicator. Highnam (1965) reviewed this topic adequately and, by the use of four diagrams, illustrated the relationship which exists in all neurosecretory cells between release, transport, and synthesis. When all three parts are in balance, the amount of hormone released will equal the amount being synthesized and may vary over a wide range, limited only by the physical and chemical dynamics of the specific cell. If the release from the parikaryon into the axon, or rate of transport down the axon, or release from the axon terminal to the vascular system is reduced, then the possibility of storage in the entire system or proximal to the point of stoppage becomes real. In this case a false indication concerning the activity of the cell in question is given to the histologists. Highnam concludes his presentation with the following two paragraphs:

> It is clear, therefore, that variations may occur in the rate of production of neurosecretory material, in its rate of release from the axon terminals, and possibly also in its speed of transport. These three variables may change independently as a result of different stimuli and the histological appearance of a neurosecretory system at any one time is the result of what is possibly a rather complex interaction between the three factors.

Consequently, by itself the histological character of a neurosecretory system can give no indication whatsoever of the output of neurosecretory factors, i.e., of the "activity" of the system. Only if the changing histology of neurosecretory systems can be related to specific developmental or physiological events, the control of which by neurosecretory factors has previously been determined by exact experimentation, can inferences be made about neurosecretory "activity" from associated histological studies. Even then, difficulties may arise when a multiplicity of neurosecretory factors are involved, or when a single neurosecretory factor is involved in the control of more than one process.

II. NONNEURAL COMPONENTS

Although neurosecretion has dominated invertebrate endocrinology recently, nonneural tissues are also of considerable importance. Phylogenetically, specialized nerve cells possessing glandular functions precede endocrine glands, and it is in the arthropods that such glands in the invertebrates become conspicuous. In Crustacea the nonneural ectoderm gives rise to the y-organ, while similar structures in insects are the prothoracic glands and corpora allata. The Crustacea obtain androgenic glands and gonads from mesoderm, but as yet no endocrine function has been associated with any endoderm tissue.

The prothoracic glands and corpora allata of insects have been studied in more detail than any other such structure. Whereas the allata are in close proximity to the neuroendocrine system by way of nerve connections through the cardiaca, the prothoracic glands are situated in the head or thorax, with muscle fibers, tracheoles, and nerves running along the gland's longitudinal axis. The parenchymal cells of both glands are epitheloid elements with interdigitating processes exposing many cells to the organ surface. One or more nucleoli are often observed close to the nuclear envelope. Each gland has abundant mitochondria, lysomelike inclusions, occasional centrioles, and distinctive Golgi bodies. A significant cytoplasmic characteristic of the two glands is the almost complete lack of any neurosecretory granules. However, both glands have axons containing abundant granules originating in neurosecretory cells of the central nervous system or structures such as the corpus cardiacum. Also significant is the presence, in varying degrees of development, of smooth and granular endoplasmic reticulum and free ribosomes. When one views these ultrastructural characteristics in light of well-established structures of vertebrate glands, similarities are found leading to the conclusion that the nonneural glands produce lipoid, and not proteinaceous hormones. Indeed this is consistent with the known chemistry of two insect hormones: ecdysone, a steroid from the prothoracic gland, and the juvenile hormone, a terpene from the corpora allata (pp. 170–173).

As yet, little information has been provided concerning the appearance, localization, transport, and release mechanisms for secretion from these glands. This stands in sharp contrast with the information available on neural glands of invertebrates.

References

Bern, H. A. (1963). The secretory neuron as a doubly specialized cell. *In* "General Physiology of Cell Specialization" (D. Mazia and A. Tyler, eds.), pp. 349–366. McGraw-Hill, New York.

Bern, H. A., and Hagadorn, I. R. (1965). Neurosecretion. *In* "Structure and Function in the Nervous System of Invertebrates" (T. H. Bullock and G. A. Horridge, eds.), Vol. 1, pp. 353–429. Freeman, San Francisco, California.

Charniaux-Cotton, H., and Kleinholz, L. H. (1964). Hormones in invertebrates other than insects. *In* "The Hormones" (G. Pincus, K. V. Thimann, and E. B. Astwood, eds.), Vol. 4, pp. 135–198. Academic Press, New York.

Dogra, G. S., and Tandan, B. K. (1964). Adaptation of certain histological techniques for *in situ* demonstration of the neuro-endocrine system of insects and other animals. *Quart. J. Microscop. Sci.* **105,** 455–466.

Gabe, M. (1966). "Neurosecretion." Pergamon Press, Oxford.

Hagadorn, I. R. (1967a). Neuroendocrine mechanisms in Invertebrates. *Neuroendocrinology (N.Y.)* **1,** 439–484.

Hagadorn, I. R. (1967b). Neurosecretory mechanisms. *In* "Invertebrate Nervous Systems" (C. A. G., Wiersma, ed.), pp. 115–124. Univ. of Chicago Press, Chicago, Illinois.

Hanström, B. (1931). Neue untersuchungen über sinnesorgane und nervensystem der crustaceen. I. *Z. Morphol. Okol. Tiere* **23,** 80–236.

Highnam, K. C. (1965). Some aspects of neurosecretion in Anthropods. *Zool. Jb. Physiol. Bd.* **71,** 558–582.

Horridge, G. A. (1968). The origins of the nervous system. *In* "The Structure and Function of Nervous Tissue" (G. H. Bourne, ed.), Vol. 1, pp. 1–31. Academic Press, New York.

Johnson, B. (1966). Ultrastructure of probable sites of release of neurosecretory materials in an insect, *Calliphora stygia* Fabr. (Diptera). *Gen. Comp. Endocrinol.* **6,** 99–108.

Scharrer, B. (1963). Neurosecretion. XIII. The ultrastructure of the corpus cardiacum of the insect *Leucophaea maderae*. *Z. Zellforsch. Mikroskp. Anat.* **60,** 761–796.

Scharrer, B. (1965). Recent progress in the study of neuroendocrine mechanisms in insects. *Archiv. Anat. Microscop. Morphol. Exptl.* **54,** 331–342.

Scharrer, E., and Brown, S. (1962). Neurosecretion in *Lumbricus terrestris*. *Gen. Comp. Endocrinol.* **2,** 1–3.

Smith, U., and Smith, D. S. (1966). Observations on the secretory processes in the corpus cardiacum of the stick insect, *Carausius morosus*. *J. Cell. Sci.* **1,** 59–66.

PART **2**

INVERTEBRATE ENDOCRINE SYSTEMS

CHAPTER **3**

CNIDARIA

Cnidaria, or Coelenterata, is the most primitive phylum and the most recent to join the list of those exhibiting signs of neurosecretion. The jellyfish, sea anemones, corals, and hydrozoans comprise this phylum, but we will direct our attention only to the last. The contribution of the Cnidaria to animal development is the gastrovascular cavity, or coelenteron; accordingly the members of this group are positioned above the sponges but below all other Metazoa.

Only members of the class Hydrozoa, a predominantly marine group with both solitary and colonial forms, have been recorded as showing signs of neurosecretion (Burnett and Diehl, 1964; Burnett et al., 1964; Lentz and Barrnett, 1965; Jha and Mackie, 1967).

HYDROZOA

A. Morphology

Only recently, substantial evidence justifies the claim that nerve cells are present in hydra. For many years this support was not available, but now, through electron microscopy, fine cellular detail has been recorded of quite distinct nerve elements.

The nervous system in this primitive organism consists principally of randomly placed bipolar or multipolar nerve cells which have only recently been classified into three types: sensory, ganglionic, and neurosecretory. The last two types are similar in size, shape, position, and cytoplasmic contents, except for the Golgi-associated, membrane-bounded, dense granules, about 1000 Å in diameter, which characterize the neurosecretory cell (Fig. 3-1). Since, however, the ganglion cell contains a pale vesicle of smaller than average diameter, it is possible that the two cells may be of the same type, but at different stages in differentiation and function. Formation of the granule in the perikaryon, transportation along the nerve extension or neurite, and release to the intercellular space are processes which are believed to be similar to those observed in higher organisms (Fig. 3-2). In addition to the presence of granules in neurons, there is a positive response to paraldehyde-fuchsin in the suspected neurosecretory cells which are most dense around the oral region and tentacle base.

With neurosecretory cells appearing in *Hydra*, the observation has been made that the appearance of nerves as glandular cells may have evolved at about the same time as, or even before the appearance of nerves as bioelectric coordinators of bodily functions. This hypothesis is supported by the lack of ultrastructural evidence of specialized synapses in any of the three nerve cells which are so characteristic of higher nervous systems. Also, neurohumoral transmission in the nervous system is apparently well developed since there is histochemical evidence for the presence of several neurohumors: epinephrine, norepinephrine, and 5-hydroxytryptamine, as well as the enzymes acetylcholinesterase and monoamine oxidase.

B. Physiology

When mature, nonbudding *Hydra littoralis* are transected below the hypostome and allowed to regenerate, changes are evident in the neurosecretory cells. The dense granules which normally abound in the vicinity of the Golgi apparatus of the parikaryon are greatly diminished within several hours following transection and have become positioned near the axon terminals. Liberation of the vesicle content from the axon is then facilitated by fusion with the plasma membrane, thereby releasing the active components into the intercellular spaces. The obvious hypothesis to be drawn from these observations is that a significant role is played by a neurosecretory substance in regulating growth and differentiation in the regenerating *Hydra* (Lentz, 1965a).

An interesting test of this hypothesis has been performed in which neurosecretory granules, isolated by differential centrifugation and

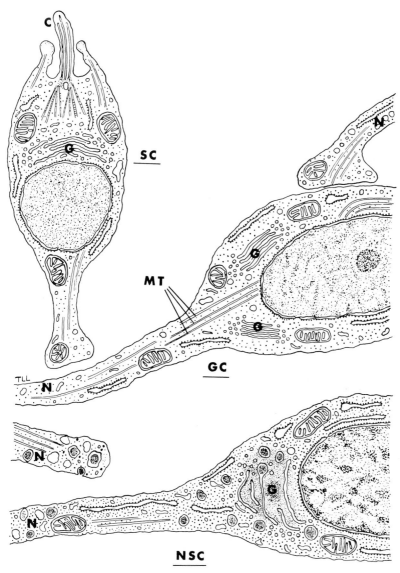

Fig. 3-1. Nerve cells in *Hydra*. SC, sensory cell; GC, ganglion cell; NSC, neuro-secretory cell. The sensory cell contains an apical cilium (C) while the opposite pole is drawn into a process terminating on a ganglion cell. The ganglion cell contains a Golgi apparatus (G), microtubules (MT), and a sparse endoplasmic reticulum. A long neurite (N) extends from the perikaryon. The neurosecretory cell is characterized by dense membrane-bounded granules in the perikaryon, especially in relation to the Golgi apparatus (G) and in the neurite. The neurites (N) contain mitochondria, ribosomes, microtubules, vesicles, or granules, and terminate in intercellular spaces, on other neurons, or on other cell types. (From Lentz, 1966.)

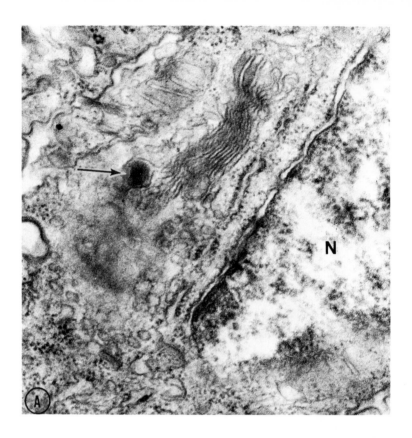

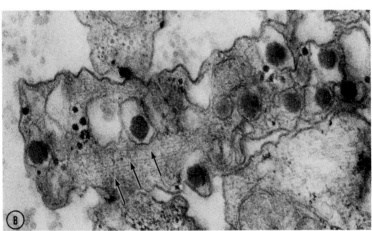

Fig. 3-2. Neurosecretory cells of *Hydra*. (A) A cell illustrating the Golgi apparatus adjacent the nucleus (N) with a neurosecretory granule present in the dilated end of a Golgi lamella (arrow). × 65,000. (B) Neurite of a cell terminating in an intercellular space containing neurosecretory granules, empty vesicles, microtubules (arrows), and dense glycogen particles. × 36,750. (From Lentz, 1966.)

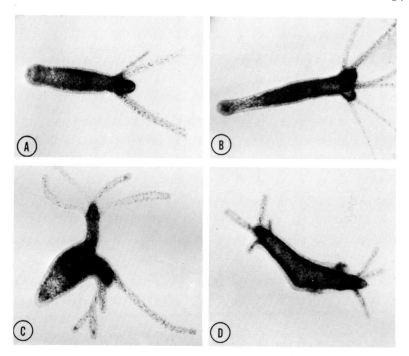

Fig. 3-3. A midsegment of *Hydra* allowed to regenerate normally (A) for 3 days is compared with three others exposed to a fraction of the centrifuged homogenate containing neurosecretory granules (B–D). The normal regenerate (A) possesses a distal head (hypostome and tentacles) and proximal base, while those exposed to the fraction containing neurosecretory granules are abnormal, having two distal heads (B), an additional head protruding from the body (C), and heads protruding both proximally and distally (D). (From Lentz, 1965b.)

checked ultrastructurally, were added to the culture media containing excised midsegments of the stomach region of *Hydra littoralis* (Lentz, 1965b). After only 4 hours in this medium, additional hypostomes and tentacles were regenerated by the body column within 3 days (Fig. 3-3). Thus, following transection, the neurosecretory cells apparently release some growth-stimulating principle, existing in a gradient between the hypostome and base, which controls cell division and regulates the form of the regenerating tissue. In the normal organism the substances may simply regulate normal growth and development.

Results have also been obtained with *Hydra pseudoligactis* which showed that similar neurosecretory cell activity occurs during normal growth, asexual reproduction, and regeneration.

Additional systems may also be under some neuroendocrine control; for example, processes containing neurosecretory vesicles have been observed in association with epitheliomuscular cells and cnidoblasts, suggesting a possible role in nematocyst discharge.

References

Burnett, A. L., and Diehl, N. A. (1964). The nervous system of Hydra. I. Types, distribution and origin of nerve elements. J. Exptl. Zool. **157,** 217–226.

Burnett, A. L., Diehl, N. A., and Diehl, F. (1964). The nervous system of Hydra. II. Control of growth and regeneration by neurosecretory cells. J. Exptl. Zool. **157,** 227–236.

Jha, R. K., and Mackie, G. O. (1967). The recognition, distribution and ultrastructure of hydrozoan nerve elements. J. Morphol. **123,** 43–61.

Lentz, T. L. (1965a). Fine structural changes in the nervous system of the regenerating hydra. J. Exptl. Zool. **159,** 181–194.

Lentz, T. L. (1965b). Hydra: Induction of supernumerary heads by isolated neurosecretory granules. Science **150,** 633–635.

Lentz, T. L. (1966). "The Cell Biology of Hydra." Wiley, New York.

Lentz, T. L., and Barrnett, R. J. (1965). Fine structure of the nervous system of Hydra. Am. Zoologist **5,** 341–356.

CHAPTER **4**

PLATYHELMINTHES

Neuroendocrine information is available for two phyla of acoelomate animals: Platyhelminthes and Nemertinea, the second of which will be discussed in Chapter 5. The flatworms are bilaterally symmetrical at some stage of their development, a characteristic which has augmented the appearance of a head and subsequent cephalization. With sensory and nervous development localized anteriorly and several crude organ systems appearing in the body, the more highly organized animal has a definite selective advantage. Thus with greater size and complexity, internal homeostatic mechanisms have evolved to control nutritive, respiratory, and excretory functions. Sense organs are only moderately developed with an equally moderate nervous system, but the important observation to be made in the case of the flatworm is that cephalization has begun with the appearance of a central nervous system controlling the organ systems (Meglitsch, 1967).

I. TURBELLARIA

A. Morphology

A centralized nervous system appears first in the flatworm with the brain as a typically bilobed ganglionic mass associated with at least one pair of ventral longitudinal nerve trunks (Fig. 4-1A). In planarians three major types of cells comprise the brain: neurons, neurosecretory,

33

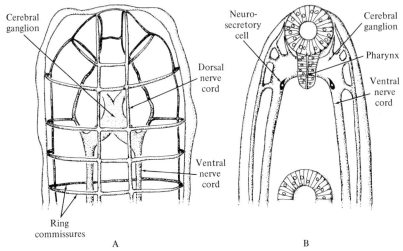

Fig. 4-1. Nervous system of Platyhelminthes. (A) Turbellaria. *Bothrioplana* with cerebral ganglion and ventral nerve cord connected to the dorsal nerve cord by way of ring commissures. (After Meglistch, 1967.) (B) Trematoda. *Dicrocoelium lanceatum* with two neurosecretory cells in the posterior region of the cerebral ganglion (After Ude, 1962.)

and neuroaccessory or neuroglial cells. Using optical microscopy, investigators have identified secretory cells in the cerebral ganglion and, in some cases, in the ventral nerve cord by their affinity to primarily the paraldehyde-fuchsin stain in the genera: *Polycelis, Dendrocoelum, Dugesia,* and *Mesostoma* (Lender, 1964; Battaglini, 1964). The cells appear to reach their greatest density just posterior to the transverse commissure. In *Mesostoma,* the presence of some accumulation of neurosecretory material has been noted in the cell bodies and within axons in the vicinity of the transverse cord which may serve as a primitive neurohemal area. The material eventually flows from the axonal extensions into the parenchyme.

Ultrastructural support for these observations has come from work on *Dugesia dorotocephala* by Morita and Best (1965). Neurosecretory granules between 400–1100 Å in diameter have been observed in many of the peripheral neurons of the brain where their formation in the perikaryon is by way of the Golgi complex (Fig. 4-2). Although the granules' diameters place them in a questionable area of the more widely accepted 1000–3000 Å range, exclusion of these cells from neurosecretory consideration on that basis alone would not be

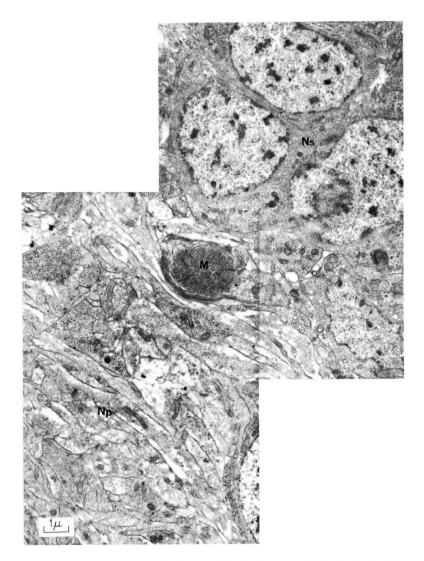

Fig. 4-2. (A) A montage of the planarian *Dugesia dorotocephala* brain showing the neuropile (Np) inside and cell bodies of nerves, and neurosecretory cells (Ns) in the periphery. A cross section of a muscle fiber (M) is designated. × 6720.

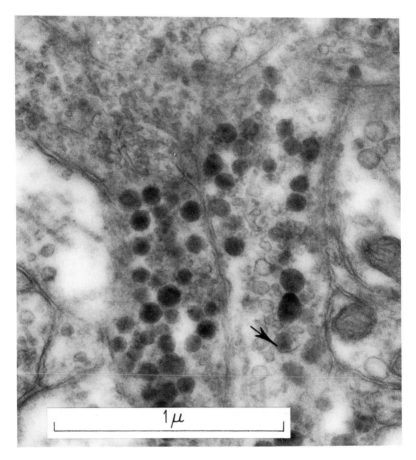

Fig. 4-2. (B) Large and small neurosecretory granules (Ng) of both ovoid and round shapes are seen in a nerve fiber with the limiting membrane of one granule evident (arrow). × 63,000. (From Morita and Best, 1965.)

justified. Following granule formation, transport is along dendritic and axonal processes to the terminal bulbs where release apparently occurs.

B. Physiology

1. REGENERATION

The relationship of the neurosecretory cells to posterior regeneration in flatworms has attracted most of the attention given to this group.

The density of neurosecretory cells in the brain of *Polycelis* increases during caudal regeneration to a maximum on the third day, while in *Dendrocoelum* a maximum is reached on the first day. Similar response is noted for the cells of the nerve cord, but with a delay of several days (Fig. 4-3). Thus, the logical observation is made that neurohormones are involved in this process. However, some doubts are cast concerning the indispensability of such cells for regeneration when the brainless posterior portion of a transected *Polycelis* can readily regenerate the head without the aid of cerebral ganglionic neurosecretory cells (Lender and Klein, 1961).

Indications also show that a substance is elaborated by the cerebral ganglion and quite possibly by neurosecretory cells which are responsible for the induction of the ocular regeneration process. However, homogenates prepared from other areas of the body as well as extracts from chick embryos induce ocular regeneration similar to that promoted by heads of planarians.

The suggestion has also been made by Best (1967) that the experiments whereby the transmission of information may take place from "trained" planarians to "untrained" ones, when trained worms are fed to the untrained, could possibly be explained on the basis of substances associated with the neurosecretory granules.

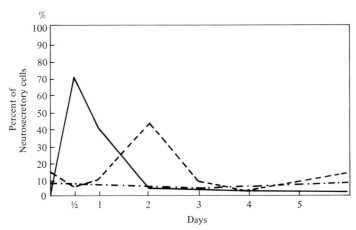

Fig. 4-3. Variations in the percentage of neurosecretory cells which appear active within the brain (—) and in the nerve trunk (– – –) in the course of 6 days during posterior regeneration. Controls (·—·—·) (From Ude, 1964.)

2. REPRODUCTION

Indications are good that the neuroendocrine system exercises some controls over reproduction in the turbellarians. Continual re-producers such as *Polycelis nigra,* and those with seasonal cycles, i.e., *Dugesia lugubris,* show a definite correlation between the number of staining neurosecretory cells and the completeness of reproductive development. During the inactive period little staining occurs in the cells or their axons, as is true of the conditions observed in organisms under fasting conditions. However, a two- to threefold increase in the number of apparent neurosecretory cells is observed in the repro-ductive planarian over that of the immature. Annual as well as daily cycles have been plotted for *Dendrocoelum lacteum* and are shown in Fig. 4-4. The maximum number of active neurosecretory cells

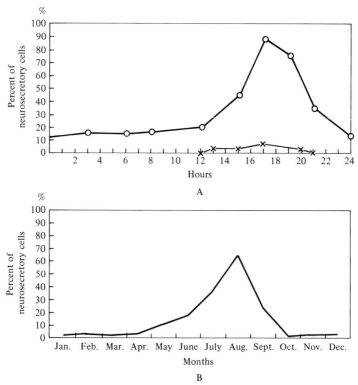

Fig. 4-4. The percentage of active neurosecretory cells in the nervous system of *Dendrocoelum lacteum* during two cycles. (A) Twenty-four-hour cycle in summer (O—O) and in winter (×—×). (B) Twelve-month cycle. (From Ude, 1964.)

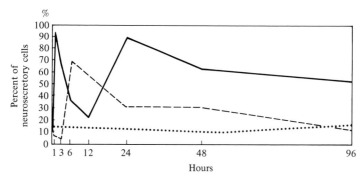

Fig. 4-5. Variations in the percentage of active neurosecretory cells in the nervous system of *Dendrocoelum lacteum* following a change in the external milieu: hypotonic (———), hypertonic (————), and control (········) (From Durchon, 1967, after Ude, 1964.)

occurs in August and a minimum of 5% extends from October to April. On the 24-hour cycle, the peak occurs in late afternoon, with the minimal number extending from midnight to 12 noon.

3. OSMOREGULATION

Ude (1964) has also shown that a response is noted following a change in the environmental osmotic pressure. The cells appear to be more responsive to the hypotonic than to the hypertonic stress concerning both time required for response and the number of cells apparently becoming stimulated (Fig. 4-5).

II. TREMATODA

Based on ultrastructural and standard histological evidence, neurosecretory cells have been identified in two genera of flukes: *Fasciola* and *Dicrocoelium*. The cell bodies are in the cerebral ganglion and are similarly arranged to the cells in the Turbellaria.

Ude (1962) shows very clearly that a single pair of secretory cells exists in *Dicrocoelium lanceatum* at the edge of the cerebral ganglion which stain positively to Gabe's paraldehydefuchsin and chrome-hematoxylin-phloxine. Fine granules of neurosecretory material appear to be evenly distributed over the cytoplasm and along the axon which extends into the brain mass (Fig. 4-1B).

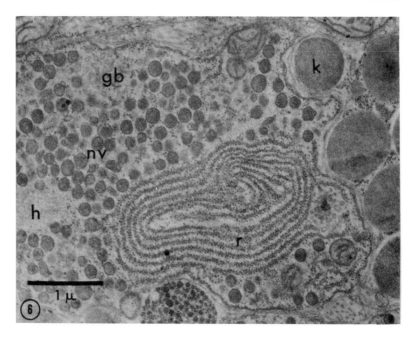

Fig. 4-6. Fine structure of a nerve cell body of *Fasciola hepatica* containing large, dense vesicles, presumably neurosecretory granules. r, endoplasmic reticulum; gb, Golgi body; k, keratin rod; h, clear area of unknown function; nv, neurosecretory granules. (From Dixon and Mercer, 1965.)

Evidence of similar neuroendocrine components in *Fasciola* is more tenuous. Dixon and Mercer (1965), in a discussion of cercarial nervous systems, grouped vesicles and granules into four classes on the bases of appearance, size, and location. One of these (type d) was possibly neurosecretory and consisted of membrane-bounded granules approximately 2000 Å in diameter with dense amorphous contents (Fig. 4-6). Although this evidence is suggestive, more support is needed on granule synthesis and transport before the ultrastructural evidence becomes as strong as that for the Turbellaria.

There has been no serious attempt to relate the limited amount of descriptive evidence of the neurosecretory system to any physiological function of the flukes.

III. CESTODA

This third flatworm class, in which the infamous tapeworms are located, has recently been reported to possess neurosecretory cells in the scolex. Employing the paraldehyde-fuchsin stain, Davey and Brechenridge (1967) observed a cluster of cells in the rostellum of *Hymenolepis diminuta*. The anterior extension of the bipolar cells forms filaments which may be sensory in nature, while the axon extends in a nerve tract from the rostellum to the lateral ganglia of the central nervous system.

A secretory cycle was observed to be associated with the development of the adult tapeworm. Fuchsinophilic material within the cell body first appears 3 days after infection, but it is 16 to 18 days before the material is observed in the axons. By that time the maximum staining of the cell body has passed, and by 40 days no fuchsinophilia is apparent in the cells. This cycle of activity may be related either to the shedding of the first proglottid or to the initiation of strobilization.

References

Battaglini, P. (1964). Attività di neurosecrezione nel sistema nervosa di un Turbellario rabdocelo: *Mesostoma linqua*. *Experientia* **20**, 150–151.

Best, J. B. (1967). The neuroanatomy of the planarian brain and some functional implications. *In* "Chemistry of Learning" (W. C. Corning and S. C. Ratner, eds.), pp. 144–164. Plenum Press, New York.

Davey, K. G., and Brechenridge, W. R. (1967). Neurosecretory cells in a cestode, *Hymenolepis diminuta*. *Science* **158**, 931–932.

Dixon, K. E., and Mercer, E. H. (1965). The fine structure of the nervous system of the cercaria of the liver fluke *Fasciola hepatica*. *J. Parasitol.* **51**, 967–976.

Durchon, M. (1967). "L'endocrinologie des Vers et des Mollusques." Masson, Paris.

Lender, T. (1964). Mise en évidence et rôle de la neurosécrétion chez les Planaires d'eau douce (Turbellaries, Triclades). *Ann. Endocrinol.* (*Paris*) **25**, Suppl., 61–65.

Lender, T., and Klein, N. (1961). Mise en évidence de cellules sécrétrices dans le cerveau de la Planaire *Polycelis nigra*. Variation de leur nombre au cours de la régénération postérieure. *Compt. Rend.* **253**, 331–333.

Meglitsch, P. A. (1967). "Invertebrate Zoology." Oxford Univ. Press, London and New York.

Morita, M., and Best, J. B. (1965). Electron microscopy studies on planaria. II. Fine structure of the neurosecretory system in the planarian *Duquesia dorotocephala*. *J. Ultrastruct. Res.* **13**, 396–408.

Ude, J. (1962). Neurosekretorische Zellen im Cerebralganglion von *Dicrocoelium lanceatum* St. u.H. (Trematoda-Digena). *Zool. Anz.* **169**, 455–457.

Ude, J. (1964). Untersuchungen zur neurosekretion dei *Dendrocoelum lacteum* Oerst. (Plathelminthes–Turbellaria) *Z. Wiss. Zool.* **170**, 233–255.

CHAPTER **5**

NEMERTINEA

This is the second of the two acoelomate phyla to have received the attention of invertebrate endocrinologists. The nemertines are similar in many respects to the flatworms and are considered just a subphylum of that group by some, while others take a more extreme view and feel that they are degenerate annelids. Nevertheless, the number of progressive features, such as a complete digestive system and a closed circulatory system, justifies their classification as a separate phylum distinct from their closest relatives, the Platyhelminthes.

The nemertines, or ribbon worms, are predominantly marine and live among algae and rocks, or in tubes constructed of mud and sand.

A. Morphology

Similarities exist between the arrangement of the nervous system in the Platyhelmenthes and nemertines. However, greater centralization has occurred in the latter, with a larger brain consisting of four ganglia (a dorsal and ventral pair) forming a nerve ring around the rhynchodaeum with the aid of dorsal and ventral commissures. A pair of lateral nerves pass from the cerebral ganglia back to the posterior end, and are there connected by an anal commissure (Fig. 5-1).

The nemertines provide the first example, phylogenetically, of a structure which possesses both glandular and nervous cells, and which is in close proximity to the vascular system; these characteristics are

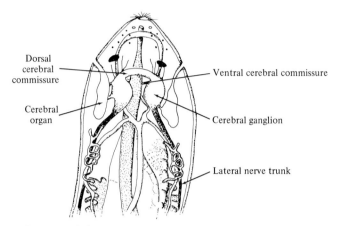

Fig. 5-1. Diagram of the anterior poi tion of the nervous system of *Amphiporus pulcher* in dorsal view showing the cerebral organs which are closely associated with the cerebral ganglion. (From Meglitsch, 1967, p. 239.)

highly suggestive of a primitive endocrine structure. The paired cerebral organs are ectodermal pits opening to the outside but closely associated with the cerebral ganglia. In the species regarded as being most highly developed, this association is more of a fusion of the cerebral organ with the cerebral ganglion (Fig. 5-2).

Scharrer (1941), on the basis of finding glandular cells containing numerous granules of an acidophile nature, suggested that this was significant evidence relating to the evolution of neurosecretion. Lechenault (1962), using chrome-hematoxylin-phloxine and paraldehyde-fuchsin, demonstrated that there were neurosecretory cells in the cerebral ganglion as well as axons extending away from the parikaryon. He was not successful in showing neurosecretion within the cerebral organ as had been suggested by Scharrer. Thus, at the present, we have evidence of neurosecretory cells in the brain, and only suggestive evidence that the cerebral organ may be the site of neurosecretory cells or the target organ of neurosecretion.

B. Physiology

Very little is known concerning the functional relationships between the observed neuroendocrine system and physiological phenomena in the nemertines. Two processes have received preliminary attention: reproduction and osmoregulation. An inhibitory control

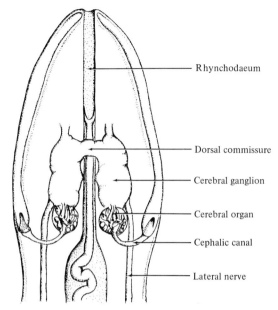

Fig. 5-2. Composite schematic representation of the cephalic region of several species of the nemeritine genus *Lineus* showing proximity of the cerebral ganglion and cerebral organ. (From Lechenault, 1962.)

appears to exist over the gonadal development in the young, for when the influence of the cerebral ganglion is altered by transection, precocious development ensues in the posterior half while similar structures of the anterior half remain nondeveloped (Bierne, 1964 and 1966).

Concerning osmoregulation, the cerebral organ and cerebral ganglion both appear to be necessary for rapid and complete adjustment in body water balance (Lechenault, 1965).

References

Bierne, J. (1964). Maturation sexuelle anticipée par décapitation de la femelle chez l'Hétéronémerte. *Lineur ruber* Müller. *Compt. Rend.* **259**, 4841–4843.

Bierne, J. (1966). Localisation dans les ganglions cérébroïdes du centre régulateur de la maturation sexuelle chez la femelle de *Lineus ruber* Müller (Heteronemertes). *Compt. Rend.* **262**, 1572–1575.

Lechenault, H. (1962). Sur l'existence de cellules neurosécrétrices dans les ganglions cérébroïdes des Lineidae (Hétéronemertes). *Compt. Rend.* **255,** 194–196.

Lechenault, H. (1965). Neurosécrétion et osmorégulation chez les Lineidae (Hétéro-rémertes). *Compt. Rend.* **261,** 4868–4871.

Meglitsch, P. A. (1967). "Invertebrate Zoology." Oxford Univ. Press, London and New York.

Scharrer, B. (1941). Neurosecretion. III. The cerebral organ of the nemerteans. *J. Comp. Neurol.* **74,** 109–130.

CHAPTER **6**

NEMATODA

Whether the reader recognizes the Nematoda as a class of either the phyla Aschelminthes or Nemathelminthes or, as in this writing, a phylum of the superphylum Aschelminthes, the material in this chapter would undergo no significant change. This is possible because, of the eight pseudocoelomate phyla, only the Nematoda has been recorded in the invertebrate neuroendocrine literature, and much of this writing has originated in Canada from Davey and his co-workers.

Nematodes, or roundworms, represent a very successful group of animals found in marine, freshwater, and terrestrial habitats. Their body is not complex and is viewed often as a system of tubes with the body wall, or cuticle, forming the outside tube. As pseudocoelomates, the nematodes possess characteristically a space, the pseudocoel, between the body wall and the digestive tube, which is filled with the perivisceral fluid. The digestive tract, the innermost tube, is a noncoiled structure which originates with the anterior mouth and terminates at a posterior anus.

A. Morphology

A circumoesophogeal nerve ring with dorsal, ventral, and lateral ganglia serves this phyla as the principal structure of the nervous system. Extending posteriorly from each ganglia are nerve cords often running the length of the body, while anteriorly smaller nerves

47

6. NEMATODA

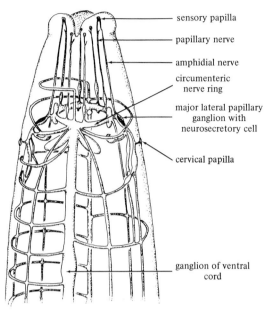

sensory papilla

papillary nerve

amphidial nerve

circumenteric
nerve ring

major lateral papillary
ganglion with
neurosecretory cell

cervical papilla

ganglion of ventral
cord

Fig. 6-1. The plan of the nervous system of *Ascaris* as viewed from the ventral side showing a neurosecretory cell within the ganglion of the major lateral papillary nerve. (Bern and Hagadorn. 1965.)

connect the peripherally located sense cells with the nerve ring (Fig. 6-1).

Only three nematodes, *Ascaris lumbricoides, Phocanema decipiens,* and *Haemonchus contortus,* have been reported as possessing neurosecretory cells. Gersch and Scheffel (1958) observed a single cell in each of the lateral ganglia of *Ascaris* which reacted positively to staining with chrome-hematoxylin as well as paraldehyde-fuchsin. These large cells varied in their staining affinity but were seldom out of phase in the same individual. Movement of stainable material, from the cell body toward the nerve ring, by way of the ventrolateral cephalic commissure, was detected. Knowledge of the number and position of observed cells in *Ascaris* was considerably enlarged by Davey (1964) when he reported twenty fuchsinophilic cells in five distinct groups. Three groups were in association with the lateral ganglia, the fourth was anterior to the nerve ring in the amphidial ganglia, and the fifth comprised most of the primary sense organs in the lip area (Fig. 6-2A). The cells in *Phocanema,* however, are in two different locations: Six cells are in the ventral ganglion

and one or two less conspicuous cells are in the dorsal ganglion (Fig.
6-2B). In *Haemonchus contortus*, neurosecretory granules (700–
1900 Å in diameter and membrane encased) appear concentrated
within axons in the vicinity of the excretory pore. Similar evidence
is observed in the ventral nerve, anterior and posterior to the excretory
duct, as well as within the ventral ganglion of the nerve ring (Rogers,
1968).

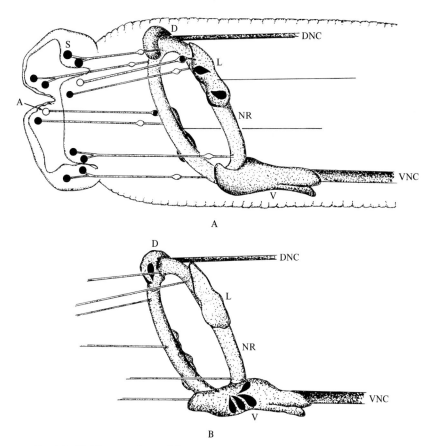

Fig. 6-2. (A) Diagram of part of the nervous system of the anterior end of a nematode
showing the sites of presumed neurosecretory cells in black. (B) Diagram of the nerve
ring and some associated structures in *Phocanema* showing the neurosecretory areas in
black. A, amphid; D, dorsal ganglion; DNC, dorsal nerve cord; L, lateral ganglion;
NR, nerve ring; S, sense cell; V, ventral ganglion; VNC, ventral nerve cord. (From
Davey, 1966.)

No neurohemal or nonneural structure, which may be homologous
to the cerebral organs of the nemertines, has yet been reported in
the nematodes.

B. Physiology

Since *Ascaris* does not survive well in culture outside its host,
functional analysis of its neuroendocrine system has not been attempted. This limitation, however, does not apply to *Phocanema*,
and recent attention has been directed toward the two interrelated
problems of cuticle formation and subsequent molting process, or
ecdysis. The former is a necessary precursor to the latter, but the
controlling mechanisms are apparently independent. In this regard,
by simple ligature experiments, it has been demonstrated that the
initiation of protein synthesis, which produces the new cuticle, is
not dependent upon the cephalic neurosecretory cells.

However, when the sequence of events leading to ecdysis is an-
alyzed, a pattern of control apparently dependent upon neurosecretion is found. When unfavorable culture medium prevents the
animals from molting even though the new cuticle has been formed,
the neurosecretory cells also fail to stain with paraldehyde-fuchsin,
which is not the case under normal conditions. Also, when extracts
of active neurosecretory cells are added to organ cultures of tissue
prepared to molt and with excretory glands exposed (having completed
its new cuticle formation), ecdysis is initiated (Davey and Kan, 1967
and 1968).

The suspected hormonal control of ecdysis in *Phocanema* appears,
then, to be of a first-order neuroendocrine reflex which can be
summarized in the following way. After the new cuticle has been
deposited, unknown factors in the culture medium initiate the synthesis and release of a hormone within the neurosecretory cells of
the nerve ring. This hormone then stimulates the single-cell excretory
gland just posterior to the nerve ring to synthesize leucine amino-
peptidase, an important constituent of molting fluid. Via the excretory
duct, the enzyme is released into the space between the old and new
cuticle with subsequent splitting of the old cuticle and ecdysis. Because of the small number of cells involved and their unique dimensions, this endocrine system offers interesting possibilities for more
detailed sudy at the ultrastructural and molecular levels on the
response of the neurosecretory cells to the environmental stimuli, as
well as the response of the excretory gland cell to neurosecretion.

References

Bern, H. A., and Hagadorn, I. R. (1965). Neurosecretion. In "Structure and Function in the Nervous System of Invertebrates" (T. H. Bullock and G. A. Horridge, eds.), Vol. 1, pp. 353–429. Freeman, San Francisco, California.

Davey, K. G. (1964). Neurosecretory cells in a nematode, Ascaris lumbricoides. Can. J. Zool. 42, 731–734.

Davey, K. G. (1966). Neurosecretion and molting in some parasitic nematodes. Am. Zoologist 6, 243–249.

Davey, K. G., and Kan, S. P. (1967). Endocrine bases for ecdysis in a parasitic nematode. Nature 214, 737–738.

Davey, K. G., and Kan, S. P. (1968). Molting in a parasitic nematode. Phocanema decipiens. IV. Ecdysis and its control. Can. J. Zool. 46, 893–898.

Gersch, M., and Scheffel, H. (1958). Sekretorisch tätige zellem in nervensystem von Ascaris. Naturwissenschaften 45, 345–346.

Rogers, W. P. (1968). Neurosecretory granules in the infective stage of Haemonchus contortus. Parasitology 58, 657–662.

CHAPTER **7**

MOLLUSCA

Consideration now moves from the preceding four lower phyla which have received scant attention from the neuroendocrinologists to the three larger and more highly evolved phyla, constituting the bulk of the literature on invertebrate endocrines. Two major evolutionary lines are apparent; The first to be discussed includes the mollusks, characterized by a lack of true segmentation. The second includes the annelid-arthropod line, characterized by segmentation. Neurosecretion will remain a key subject of our consideration even though the emergence of pure endocrine organs will be observed.

Neurosecretion in the Mollusca was first proclaimed by Berta Scharrer (1935). This discovery was, in fact, the beginning of her immensely productive efforts in the field of neuroendocrinology. Since that time over a hundred molluskan species have been reported to exhibit some signs of neuron secretion, with approximately 75% of the total representing the class Gastropoda (Durchon, 1967). An interesting endocrine relationship, and one which is quite popular with textbook writers, appears in the Cephalopoda where the optic glands influence sexual maturation (Fig. 7-13). However, in the same class, and to the amazement of many investigators, the search for neurosecretory cells until quite recently has been a rather unrewarding endeavor.

In light of improved histological techniques and subsequently modified guidelines for evaluating potential neurosecretory systems, a justifiable reinterpretation or critical assessment of the gastropod literature is found in the paper by Simpson, Bern, and Nishoika

53

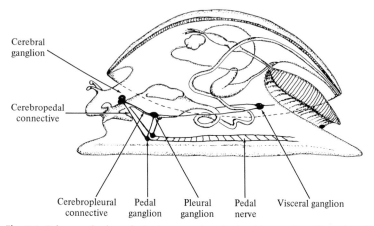

Cerebral
ganglion

Cerebropedal
connective

Cerebropleural   Pedal   Pleural   Pedal   Visceral ganglion
connective   ganglion   ganglion   nerve

Fig. 7-1. Scheme of a hypothetical ancestral mollusk with attention directed to the nervous system and four principal ganglia. (From Meglitsch, 1967, p. 458.)

(1966). It appears to this writer that the above-mentioned work, which was part of a symposium on invertebrate endocrinology, might well become the starting point for new research into molluskan neuroendocrinology, and, in fact, would assist those interested in any invertebrate neuroendocrine system.

Of the five classes of Mollusca which have attracted the attention of the researchers on neurosecretion, Amphineura has brief mention by Scharrer (1937), along with a recent report of active cells appearing in the buccal ganglia of three species of chitons. Only one species of the Scaphopoda has been studied, and that by Gabe (1949). However, the Gastropoda, Pelecypoda, and Cephalopoda have been studied by numerous investigators and will provide the material for this chapter.

Although mollusks are found in diverse habitats, there are general morphological features of the nervous system common to all. The hypothetical ancestral nervous system has, as its center, three pair of ganglia: cerebral, pedal, and pleural (Fig. 7-1). The cerebral pair, or supraesophageal ganglia, lies above and anterior to the esophagus, the pedal pair, below the esophagus, and the pleural, posterior and lateral to the esophagus. Connectives link the three ganglia, thus forming a neural triangle on each side of the gut, and commissures connect the members of each pair. Often a fourth pair, the visceral ganglia, are connected to the pleural ganglia by way of the visceral nerves.

I. GASTROPODS

In snails, a definite centralized nervous system, composed principally of the triangulated cerebral, pleural, and pedal ganglia, lies near the esophagus. A pair of nerves leads away from each mass and forms, near their terminus, a pair of ganglia. The exception is the pedal nerve trunk which forms no ganglion as it extends into the foot. The cerebral pair gives rise to the buccal ganglia, and from the pleural appears the parietal, then the visceral ganglia. Owing to body torsion, the positions of the visceral and parietal ganglia are often distorted, thereby losing symmetry in organization; the pulmonates, however, regain bilateral symmetry by shortening the visceral nerves. This modification may become so extreme in some cases that the visceral and pleural ganglia unite with the cerebral ganglia to form a complex brain with peripheral nerves extending to the various body regions (Fig. 7-2).

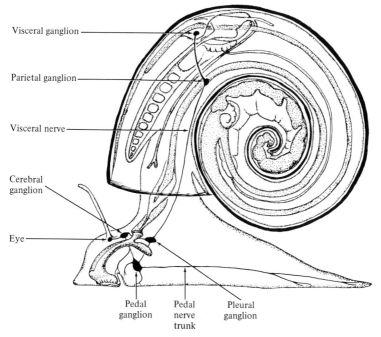

Fig. 7-2. Generalized scheme of organization of the nervous system of a gastropod. Twisting and shortening of the visceral nerve plus fusion of several ganglia are found in various groups. (From Meglitsch, 1967, p. 506.)

Cells which have shown indications of being neurosecretory are rather ubiquitous in gastropods, having been recorded in all orders and in most portions of the central nervous system of one species or another.

It is largely to this problem that Simpson, Bern, and Nishioka (1966) directed their comments; they called for a more deliberate approach to the discovery and evaluation of neurosecretory cells. A slightly modified view is presented by Boer, Douma, and Koksma (1968), justifying histochemical and electron microscope techniques for the designation of neurosecretory components in Mollusca.

Several neuroglandular organs have been tentatively described on the basis of the appearance of presumably neurosecretory cells in conjunction with the sensory or glandular cells. The most valid examples of this are probably the mediodorsal bodies in pulmonates.

A. Prosobranchia

Twenty-five species from this subclass were recorded by Gabe (1953) to exhibit neurosecretory cells in all principal ganglia of one species or another except the buccal and pedal, which do not stain at all. These were small to medium cells with the greatest diameter between 10 to 20 μ. Transport of material from the perikaryon into their axons occurred, but difficulty has been encountered in mapping the paths of such axons.

Indirect evidence, provided by some histophysiological studies, indicates that certain groups of neurosecretory cells of the prosobranchs follow a cyclical pattern of synthesis and release which can be related to sexual maturation. A generalized scheme might be that the reproductively mature, fully developed animal has less neurosecretory material in the parikaryon of suspected cells than the same organism under a period of sexual repose. An alteration in the number of cells staining has also been noted in *Vivipara vivipara* with a maximum occurring during the early summer months. The interpretation may be that, as in other invertebrates, active neurosecretory cells will not stain heavily, since the synthesized material is immediately utilized in regulating physiological mechanisms. When the demand for the hormone is then reduced or entirely eliminated, the cells may come to rest either devoid of stainable material or retaining a certain residual amount.

B. Opisthobranchia

Large specialized secretory cells are found with greater frequency and in larger numbers in the cerebral and pleural ganglia than in other ganglia of this subclass. Ultrastructural examination of the suspected cells has indicated that osmophilic, membrane-limited granules, of the size characteristic of elementary neurosecretory granules, are present in the perikarya and axons.

In association with the posterior aspect of the cerebral ganglia, a structure with apparent endocrine importance, termed organe juxtaganglionnaire by Martoja (1965), has been found in the opistho-branches. Its exact position as well as its relationship to adjacent tissue varies in different species. In a review of the neurosecretory phenomena in this subclass, Vicente (1966) indicates that the organ's main function may be similar to that of the sinus gland of the Crustacea, that is, to perform as a neurohemal structure. Histo-chemical analyses appear to indicate that the secretion from this gland seems to be proteinaceous, and not lipoid in nature.

Two physiological processes appear to be responsive to hormonal control in the opisthobranches: osmoregulation and reproduction. When the pleural ganglia are removed, a loss of weight occurs in the organism which is quickly recovered upon the reimplantation of normal ganglia. The possibility of an antidiuretic factor originating in these ganglia is thus suggested.

More can be said of reproduction, the second system which re-sponds to gland manipulations. As summarized by Vicente, neuro-secretory cells in the cerebral ganglia apparently synthesize and release a gonadotropic hormone which is passed via axons to the organe juxtaganglionnaire, and stored there (Fig. 7-3). If the cerebral ganglia are removed, mating and egg laying are suppressed and the removal of other ganglia seem to modify these conditions in no way. During active gametogenesis the organe juxtaganglionnaire is well developed, but it reduces its size upon the release of the sexual products.

On the other hand, neurosecretory cells located in ganglia at the base of modified head tentacles (rhinophores) are believed to secrete a hormone which inhibits reproductive development. Thus, in sexual repose a balance exists between the gonadotropic and inhibitory hormone. At the beginning of the reproductive season the inhibiting hormone is suppressed and the gonodotropic hormone dominates.

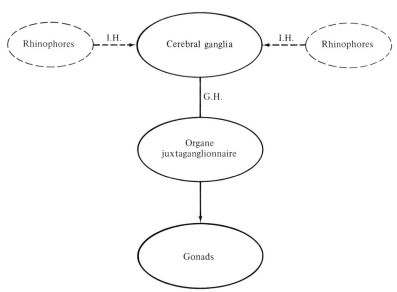

Fig. 7-3. The suggested mechanism for the control of reproduction in the Opistho-branchia. The gonadotropic hormone (G.H.) synthesized by the neurosecretory cells of the cerebral ganglia is passed to the organe juxtaganglionnaire for storage before acting on the gonads. Prior to sexual maturity or during periods of sexual rest, the in-hibiting hormone (I.H.) suppresses the G.H., probably at the site of synthesis.

With ablation of the rhinophore, the organe juxtaganglionnaire, no longer being held in check by the inhibiting hormone, shows con-siderable hypertrophy.

If these hypothetical mechanisms stand the tests of further experi-mentation, then the above is a physiological process under the con-trol of two opposing hormones; this characteristic will be repeated often as we move into the higher phyla.

It is possible that the organe juxtaganglionnaire will prove to be more than simply a neurohemal organ as stated by Vicente, for it may function as an endocrine organ in its own right with intrinsic secretory cells. If this is the case, then the diagram in Fig. 7-3 may illustrate a second order neuroendocrine reflex with neurosecretory cells controlling the organe juxtaganglionnaire, which then controls the gonads.

C. Pulmonata

The pulmonates have attracted more research activity than either of the two previously discussed subclasses. It would be incorrect to state that a better understanding exists as a result of this activity, for such certainly does not appear to be the case. A number of apparent contradictions exist which, hopefully, will be unraveled with time. The previously mentioned critical discussion by Simpson et al. (1966) will certainly help in this direction, and the study of pulmonates will undoubtedly benefit from this work more than the study of the other gastropods.

Areas of the central nervous system with the greatest neurosecretory activity are the cerebral, pleural, parietal, and visceral ganglia. From the work of Lever and others (1957 and 1961), location of the neurosecretory cells in Lymnaea stagnalis can be pinpointed as shown in Fig. 7-4A, which not only demonstrates the location of the neurosecretory cells but shows the extent of ganglia contraction in the more advanced gastropods. Several reports have classified these cells in two to five types, based on the morphological characteristics observed following differential staining. However, drawing on the extensive research from the Free University in Amsterdam, Boer and associates have recently brought up to date the problem concerning neurosecretory cells in Lymnaea stagnalis. As illustrated in Fig. 7-4B, there are two cell types in three locations in the cerebral ganglion. Two groups, with approximately fifty cells each, appear to be histochemically similar, for they stain well with chrome-hematoxylin and paraldehyde-fuchsin (Gomori-positive cells), and are labeled mediodorsal (MDC) and laterodorsal (LDC), which reflects only their differing location within the ganglion (Fig. 7-5A). The third group, located medially from the LDC but shown in the figure as positioned between the previous two groups, contains the caudodorsal cells (CDC) (Fig. 7-5B). These cells are histochemically different from the MDC and LDC, for their neurosecretory product is phloxinophilic (Gomori-negative cells).

Axons leading from the three groups of cells have been traced and potential neurohemal areas have been identified. The MDC and LDC axons tend to group into a single bundle which passes out of the ganglion and into the median lip nerve. The neurosecretory product is then stored in bulb-shaped axon endings just under the perineurium (Fig. 7-6A and B). Axons from both the right and left CDC

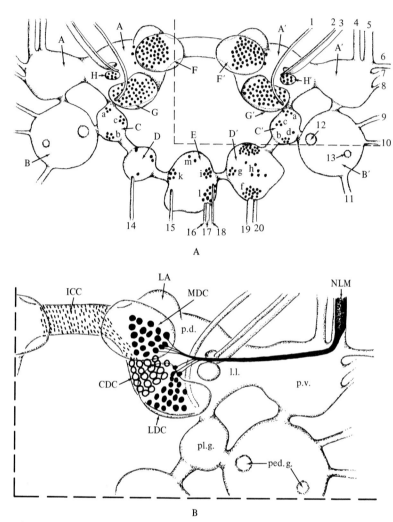

A

B

Fig. 7-4. A. Location of "Gomori-positive" cells and cell groups in central ganglia of *Lymnaea stagnalis*. Ventral parts of cerebral ganglia and pedal ganglia are turned to the sides. A,A', cerebral ganglia; B,B', pedal ganglia; C,C', pleural ganglia; D,D', parietal ganglia; E, visceral ganglia; F,F', mediodorsal bodies; G,G', laterodorsal bodies; H,H', lateral lobes, a–m, groups of "Gomori-positive" cells; 1–20, nerves (From Lever *et al.*, 1961). B. Schematic drawing of the neurosecretory system of the cerebral ganglion. CDC, caudodorsal cells; ICC, intercerebral commissure; LA, lobus anterior; LDC, laterodorsal cells; MDC, mediodorsal cells; NLM, nervus labialis medius; l.l., lateral lobe; p.d., pars dorsalis of the cerebral ganglion; p.v., pars ventralis of the cerebral ganglion; pl.g., pleural ganglion; ped. g., pedal ganglion. (From Boer, Slot, and vanAndel, 1968.)

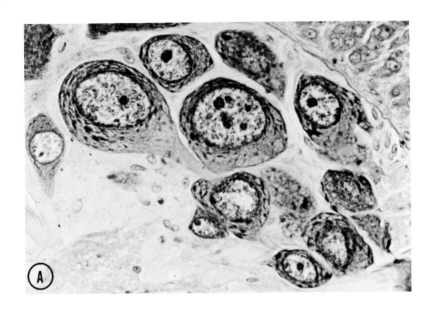

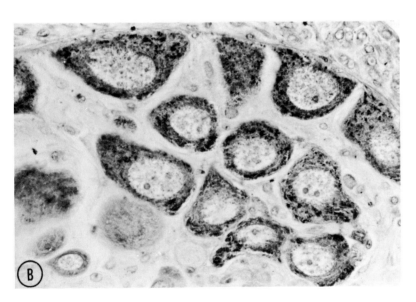

Fig. 7-5. Mediodorsal or laterodorsal neurosecretory cells of *Lymnaea stagnalis* during winter inactivity with only a small amount of neurosecretory material present in the cytoplasm (× 455). B. A group of caudodorsal neuronic cells with phloxinophilic masses between the Nissl-disks (× 455). (From Joosse, 1964.)

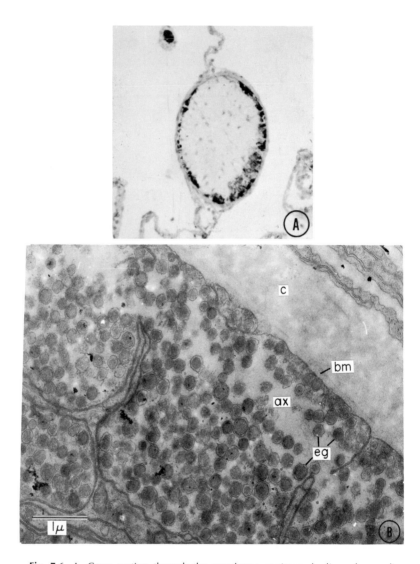

Fig. 7-6. A. Cross section through the membrana capitocerebralis and a median lip nerve from *Lymnaea stagnalis* (×175). (From Joosse, 1964.) B. Peripheral part of the median lip nerve showing bulb-shaped endings (ax) filled with elementary granules (eg); c. nerve capsule; bm, basement membrane. (From Boer, Douma, and Koksma, 1968.)

group of cells pass into the intercerebral commissure and abut on the peripheral surface (Fig. 7-4).

The ultrastructure of the MDC and LDC is similar to the typical neurosecretory cell with electron-dense, membrane-bound, elementary granules averaging 2000 Å in diameter, as well as the normal organelles of mitochondria, rough endoplasmic reticulum, ribosomes, and Golgi bodies. No cytoplasmic differences can be detected between these two cell groups, or between cells of different sizes within these groups. CDC, on the other hand, have fewer elementary granules which average only 1500 Å in diameter and are more electron-dense than the granules in the MDC and LDC.

Although no conclusive evidence has been offered as to the functional importance of any of these cells, there is an apparent annual cycle with peak cellular activity occurring in the spring (Joosse, 1964). Active cells have been found to contain a lower density of elementary granules than inactive cells.

Attached to the cerebral ganglia in several pulmonates are two structures which have received divided functional interpretation within the neuroendocrine framework (Fig. 7-7). First, the medio-

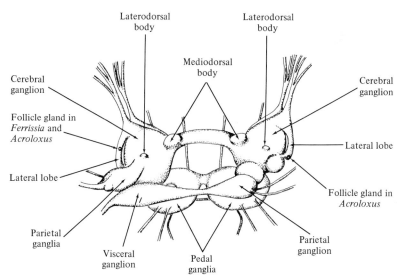

Fig. 7-7. Central nervous system of *Ferrissia* (Pulmonata) illustrating the compact nature of the ganglia plus the mediodorsal bodies and follicle glands which may possess some neuroendocrine function. (From Bern and Hagadorn, 1965, after Lever, 1957, 1958.)

dorsal bodies are small paired structures which may possess some structural relationship with the MDC, LDC, or CDC neurosecretory cells just below in the cerebral ganglion. A hypothesis which was supported by Joose (1964) is that they are endocrine organs and are involved in some way with the processes of ovulation and oviposition. This has received further support by Boer, Slot, and vanAndel (1968), but the question still remains unsettled. It has also been suggested that the bodies are involved in the control of osmoregulation as well as reproduction. The second structure, the paired follicle gland (cerebral gland), is much smaller and is attached to the lateral lobes, arises from the cerebral ganglia. This gland has also caused conflicting speculation as to whether it is or is not an integral part of the neuro-endocrine system. Simpson et al., (1966) stated that it ". . . is best placed in the category of vesicular structures with unknown func-tion. . . ." The status of this tissue is still questionable even after Brink and Boer (1967) brought electron microscopy to bear on the problem.

In 1961 Pelluet and Lane reported an interesting study in which egg production of two species of the slug, *Arion*, could be drastically modified through experimental procedures. By extirpating the optic tentacles, a significant increase was observed in the production of eggs by the ovotestis which could then be returned to the control level by injecting an extract of the removed tentacles. Also the injection of the brain extract (composed of the five pairs of ganglia) into intact animals stimulated egg production, but no change followed the in-jection of tentacular extract. These investigators then postulated a dual hormonal control mechanism with the tentacle hormone regulating sperm production and the brain hormone controlling egg production. Further work on this subject has attempted to classify the cells around the tentacle base according to their neurosecretory potential.

From experimental work on snails, evidence is also present which suggests that two additional physiological processes might be under endocrine control. First, in the cerebral ganglion a relationship exists between neurosecretory cells of the lateral lobe and osmoregulation. Second, hibernation may be under the control of an active sub-stance secreted from the visceral ganglia.

II. PELECYPODS

The sedentary habits of the bivalved mollusks (clams, oysters, mussels, and scallops) are undoubtedly reflected in the simplicity

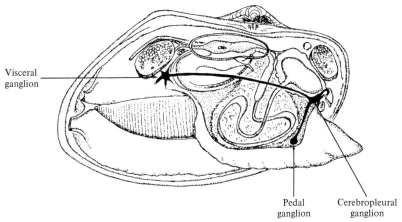

Visceral
ganglion

Pedal   Cerebropleural
ganglion   ganglion

Fig. 7-8. Generalized scheme of the organization of the nervous system of a pelecy-
pod. The cerebral and pleural ganglia have united and the parietal ganglia have been
lost. (From Meglitsch, 1967, p. 535.)

of their nervous systems. The cerebral and pleural ganglia of the
neural triangle have united in all orders except the most primitive
Protobranchia, with only a single pair of connectives extending to
the pedal ganglia. No parietal ganglia are present, so the cerebropleural are connected directly to the paired visceral ganglia which
are joined by a commissure (Fig. 7-8).

Neurosecretory cells have been commonly observed in the cerebropleural, in the visceral, and, less frequently, in the pedal ganglia.
Considerable variation exists concerning the density and location
of neurosecretory cells which seldom exceed 20 μ in their maximum
diameter. An exception to this occurs in *Dreissena polymorpha*
where the cells of the largest of three groups of neurosecretory cells
sometimes reach 60 μ in length, with the nuclear diameter between
10–15 μ.

From a functional point of view, Lubet (1955, 1956) was the first
to suggest that a relationship might exist between the neurosecretory
cells in the cerebropleural and visceral ganglia and reproduction in
two marine mussels, *Mytilus edulis* and *Chlamys varia*. During the
period of gamete maturation he found that cerebral neurosecretory
material accumulates within the perikaryon and is discharged just
prior to the extrusion of gametes. This pattern of cell synthesis and
release is repeated before each subsequent gamete evacuation.

Surgical removal of the cerebral ganglia has little effect on maturation of the gonocytes, but accelerates their discharge. Lubet thus concluded that the cerebropleural ganglia in some way inhibited spawning until a few days before the beginning of the reproductive period, at which time this inhibition was removed coincident with the discharge of neurosecretory substances. The transport of this material may be via axons directly to a target site, thus insuring stimulation of the correct structure at the appropriate time. The mussel then becomes sensitive to the correct environmental stimuli with completion of the spawning process.

An analogous study of the freshwater mussel, *Dreissena polymorpha*, illustrates that quite the opposite relationship exists in a similar organism. Antheunisse (1963) concluded that even though parallels existed between the neurosecretory and reproductive cycles, the cerebropleural ganglia did not exercise any influence over the growth of the oocytes, or any control over spawning.

III. CEPHALOPODS

On the basis of complexity of structure and function, the octopi and squids have the most highly developed nervous system to be found in invertebrates. The cerebral, pedal, pleural, and visceral ganglia can still be recognized, but extensive fusion and restructuring have altered the plan of the generalized mollusk (Fig. 7-9). The cerebral ganglia are now referred to as the supraesophageal mass while the pedal, pleural, and visceral ganglia comprise the subesophageal region of the brain. The cephalopods are apparently more independent of a neurosecretory final common pathway and, instead, utilize the central nervous system to control gonadotropic hormone production. The transduction or integration of the environmental signals still must occur in some nerve cell(s), but the response is apparently in the form of a bioelectric potential rather than a secretion.

Even in the face of this current lack of evidence for neurosecretion in the cephalopod cerebral ganglion, there are several noteworthy reports on the presence of neurosecretion elsewhere. Barber (1967) reported on the presence of neurons and axons containing membrane-bound, electron-dense granules ranging from 1000–2000 Å in diameter in the juxtaganglionic tissue of *Octopus vulgaris*. This nervous tissue is associated with the inferior buccal ganglia which, together

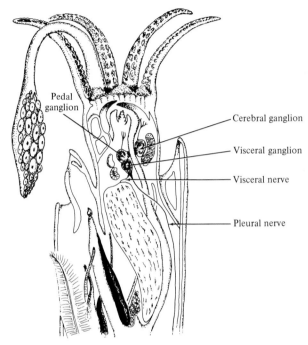

Fig. 7-9. The organization of the nervous system of the common squid, *Loligo*. After considerable fusion only three ganglionic masses are apparent: cerebral, pedal, and visceral ganglia. (From Meglitsch, 1967, p. 556.)

with the superior buccal ganglia, make up the buccal complex; this complex, in turn, may be partly fused with the cerebral ganglion. The axons pass to the surface of the buccal sinus which appears to serve as the release area for the neurosecretion (Fig. 7-10). Alexandrowicz (1964, 1965) reported that neurosecretory cells originating in the visceral ganglion of the subesophageal portion of the brain extend posteriorally along the vena cava, in close association with several visceral nerves. This work has recently been supported with ultrastructural observations by Martin (1968). The functional significance of this system may prove to be quite similar to that of the pericardial tissue of crustaceans in exerting some control over the circulatory system.

Sexual maturity in the female octopus is regulated by secretions of the optic gland, a small body lying on each of the two optic stalks. This gland is composed of stellate glandular cells within a connective

68

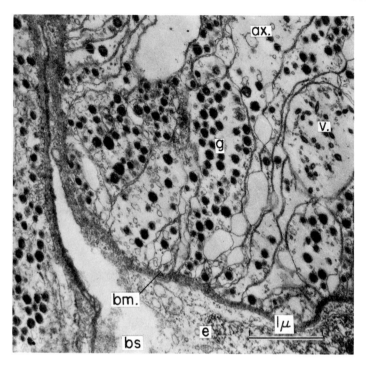

Fig. 7-10. Axons (ax) of the juxtaganglionic tissue of *Octopus vulgaris*, ending on the basement membrane (bm) of the buccal sinus (bs). Electron-dense granules (g), as well as the less common clear vesicles (v), are also present. e, Endothelial cell. (From Barber, 1967.)

tissue network and, when active, the cells experience an increase in size with the formation of vesicles containing a yellowish secretion. This is then taken up by an abundant blood supply which enters the gland from the cephalic artery. The production of this hormone is regulated by nerves originating in the small bilateral subpedunculate lobe in the supraesophageal portion of the brain, which in turn is innervated by the optic nerves.

The relationship between these structures and reproduction is illustrated in Fig. 7-11, which shows the well-known experiments performed by Wells and Wells (1959; Wells, 1960). Interruption of the nervous signal by transection or ablation at any of four locations (optic nerves, base of optic lobe, optic tract, or subpedunculate lobe) will apparently remove the gland from an inhibitory control and allow it to enlarge, leading to subsequent gonadal development. However, when the removal of the optic gland accompanies these procedures, ovarian enlargement does not occur. On the strength of these and other observations it has been concluded that an environment signal, probably some change in the photoperiod, generates a signal in the peripheral visual pathway which inhibits the full development of the optic gland by way of the subpedunculate lobe. Once such inhibition over the optic gland has been removed under natural conditions or experimentally, the gland enlarges and releases the gonadotropic factor, which is followed by the enlargement of the ovary and, consequently, the completion of the reproductive cycle. There is, nevertheless, a certain competence or responsiveness which must exist in the target organ before the ovaries are affected by the presence of the gonadotropic hormone.

Similar experiments have much the same response in the male

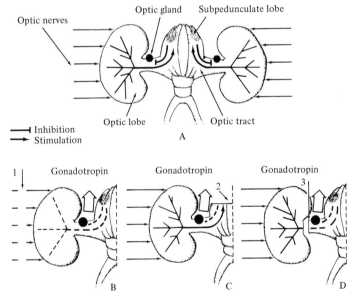

Fig. 7-11. Neuroendocrine control of gonad maturation in the octopus. A. In an immature animal, unoperated upon, the production of gonadotropin by the optic glands is held in check by an inhibitory nerve supply originating in the subpedunculate lobe of the brain. Activation of this brain center appears to depend upon changes in photoperiod. B. Section of the optic nerves (point 1) prevents activation (broken lines) of the inhibitory nerve center; the optic glands enlarge and secrete gonadotropin, which induces hypertrophy of the gonads. The same result may be accomplished by ablation of the subpedunculate lobes (C, at point 2) or by section of the optic tract (D, point 3). (From Turner, 1966, after Wells and Wells, 1959.)

octopus. The ablation of the subpedunculate lobe in a young male causes an enlargement of the optic gland, the appearance of spermatophores, and an increase in weight of testes comparable to that in control animals.

Thus, the endocrinology of reproduction in the cephalopods, the most advanced mollusks, involves a control mechanism which excludes all signs of neurosecretion, but has evolved an endocrine organ comparable in some ways to the corpus allatum in insects. In both cases, the octopi and insects, the secretion of a gonadotropic hormone is prevented by inhibitory innervation.

References

Alexandrowicz, J. S. (1964). The neurosecretory system of the vena cava in Cephalopoda. I. *Eledone cirrosa. J. Marine Biol. Ass. U. K.* **44,** 111–132.
Alexandrowicz, J. S. (1965). The neurosecretory system of the vena cava in Cephalopoda. II. *Sepia officinalis* and *Octopus vulgaris. J. Marine Biol. Ass. U. K.* **45,** 209–228.
Antheunisse, L. J. (1963). Neurosecretory phenomena in the zebra mussel *Dreissena polymorpha* Pallas. *Arch. Neerl. Zool.* **15,** 237–314.
Barber, V. C. (1967). A neurosecretory tissue in *Octopus. Nature* **213,** 1042–1043.
Bern, H. A., and Hagadorn, I. R. (1965). Neurosecretion. *In* "Structure and Function in the Nervous System of Invertebrates" (T. H. Bullock and G. A. Horridge, eds.), Vol. 1, pp. 353–429. Freeman, San Francisco, California.
Boer, H. H., Douma, E., and Koksma, J. M. A. (1968). Electron microscope study of neurosecretory cells and neurohaemal organs in the pond snail *Lymnaea stagnalis. Symp. Zool. Soc. London* **22,** 237–256.
Boer, H. H., Slot, J. J., and vanAndel, J. (1968). Electron microscopical and histochemical observations on the relation between medio-dorsal bodies and neurosecretory cells in the Basommatophoran Snails *Lymnaea stagnalis, Ancylus fluviatilis, Australorbis glabratus* and *Planorbarius corneus. Z. Zellforsch. Mikroskop. Anat.* **87,** 435–450.
Brink, M., and Boer, H. H. (1967). An electron microscopical investigation of the follicle gland (cerebral gland) and of some neurosecretory cells in the lateral lobe of the cerebral ganglion of the Pulmonate Gastropod *Lymnaea stagnalis* L. *Z. Zellforsch. Mikroskop. Anat.* **79,** 230–243.
Durchon, M. (1967). "L'endocrinologie des Vers et des Mollusques." Masson, Paris.
Gabe, M. (1949). Sur la présence de cellules neurosécretrices chez *Dentalium entale* Deshayes. *Compt. Rend.* **229,** 1172–1173.
Gabe, M. (1953). Particularites morphologiques des cellules neurosécretrices chez quelques Prosobranches monotocardes. *Compt. Rend.* **236,** 323–325.
Joosse, J. (1964). Dorsal bodies and dorsal neurosecretory cells of the cerebral ganglia of *Lymnaea stagnalis* L. *Arch. Neerl. Zool.* **16,** 1–103.
Lever, J. (1957). Some remarks on neurosecretory phenomena in *Ferrissia* sp. (Gastropoda Pulmonata). *Koninkl. Ned. Acad. Wetenschap., Proc.* (C) **60,** 510–522.
Lever, J. (1958). On the occurrence of a paired follicle gland in the lateral lobes of the cerebral ganglia of some Ancylidae. *Proc. Kon. Ned. Akad. Wet.,* (C) **61,** 235–242.

Lever, J., Kok, M., Meuleman, E. A., and Joosse, J. (1961). On the location of Gomori-positive neurosecretory cells in the central ganglia of *Lymnaea stagnalis*. *Koninkl. Ned. Acad. Wetenschap., Proc. (C)* **64**, 640–647.

Lubet, P. (1955). Le determinisme de la ponte chez les lamellibranches (*Mytilus edulis* L.). Intervention des ganglions nerveux. *Compt. Rend.* **241**, 254–256.

Lubet, P. (1956). Effets de l'ablation des centres nerveux sur l'émission des gamètes chez *Mytilus edulis* L. et *Chlamys varia* L. (Mollusques Lamellibranches.) *Ann. Sci. Nat. Zool. Biol. Animal* [11] **18**, 175–184.

Martin, R. (1968). Fine structure of the neurosecretory system of the vena cava in *Octopus*. *Brain Research.* **8**, 201–205.

Martoja, M. (1965). Existence d'un organe juxta-ganglionnaire chez *Aplysia punctata* Cuv. (Gasteropode Opisthobranche.) *Compt. Rend.* **260**, 4615–4617.

Meglitsch, P. A. (1967). "Invertebrate Zoology." Oxford Univ. Press, London and New York.

Pelluet, D., and Lane, N. J. (1961). The relation between neurosecretion and cell differentiation in the ovotestis of slugs (Gasteropoda: Pulmonata). *Can. J. Zool.* **39**, 789–805.

Scharrer, B. (1935). Ueber das Hanströmsche Organ X bei Opisthobranchiern. *Pubbl. Staz. Zool. Napoli* **15**, 132–142.

Scharrer, B. (1937). Uber sekretorische tätige Nervenzellen bei wirbellosen Tieren. *Naturwissenschaften* **25**, 131–138.

Simpson, L., Bern, H. A., and Nishioka, R. S. (1966). Survey of evidence for neurosecretion in gastropod molluscs. *Am. Zoologist* **6**, 123–138.

Turner, C. D. (1966). "General Endocrinology." Saunders, Philadelphia, Pennsylvania.

Vicente, N. (1966). Sur les phénomènes neurosécrétoires chez les Gastéropodes Opisthobranches. *Compt. Rend.* **263**, 382–385.

Wells, M. J. (1960). Optic glands and the ovary of *Octopus*. *Symp. Zool. Soc. London* **2**, 87–108.

Wells, M. J., and Wells, J. (1959). Hormonal control of sexual maturity in *Octopus*. *J. Exptl. Biol.* **36**, 1–33.

CHAPTER **8**

ANNELIDA

A very important step in the evolution of the animal kingdom occurred with the appearance of a segmented, coelomate organism with a straight digestive tube. From this stem developed the Annelida and Arthropoda. This chapter will discuss the former, and the remainder of the book will deal with the latter.

The phylum Annelida can be divided into the following three classes (numbers in parentheses indicate the number of species of each class in which neurosecretory cells have been studied histologically as reported by Gabe, 1966): the predominately marine Polychaeta (38), the freshwater and terrestrial Oligochaeta (12), and the leeches, Hirudinea (5). This illustrates that the marine environment of polychaetes has apparently not presented a barrier to the student of annelid neuroendocrinology.

Although considerably smaller in number than the preceding phylum (Mollusca, approximately 80,000 species), the annelids, with an estimated 7000 known species, have attracted a substantial share of the invertebrate research emphasis. Beginning in the late 1930's with Berta Scharrer's report on *Nereis virens,* and extending to the present, this phyla has offered much to our understanding of the morphological and physiological implications of neurosecretion. One of the most profound findings which has influenced all levels of neuroendocrinology was that of E. Scharrer and Brown (1961) in which they presented ultrastructural evidence from the earthworm, *Lumbricus terrestis,* supporting the hypothesis that elementary neurosecretory granules are formed within the Golgi complex (Fig. 2-1).

Similar evidence has since eminated from many laboratories using vertebrate as well as invertebrate material.

Annelids with a coelom and metameric organization are also characterized by a nervous system consisting of a circumenteric nerve ring and a pair of ventral nerve cords. The nerve ring is composed of two dorsal cerebral or supraesophageal ganglia united, via a pair of connectives, with the subesophageal ganglia, which are the beginning of ventral nerve cords. Evolutionary modifications of this system in the form of lateral and longitudinal compressions have occurred, as in other phyla, to produce a more compact and probably more efficient functional unit. The lateral compression occurs with a shortening of the intrasegmental ganglionic connectives to produce, first, a union of the paired ganglia and, second, the enclosure of the two nerve cords within a single sheath. Longitudinal compression provides a class characteristic because the polychaetes retain their cerebral ganglia in the prostomium, but the oligochaetes show posterior migration of the cerebral ganglia and fusion of several ventral ganglia with the subesophageal ganglia (Fig. 8-1). The leeches

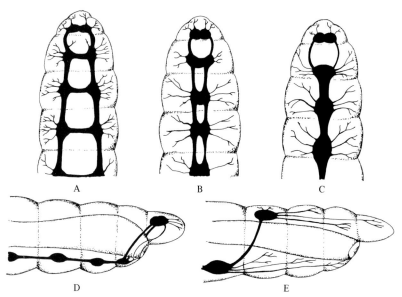

Fig. 8-1. Evolutionary tendencies observed in the annelid nervous system. A, B, and C illustrate lateral compression of the nerve cords, while E shows longitudinal shifting which has occurred in the oligochaetes and leeches as compared to the more primitive polychaetes, D. (From Meglitsch, 1967, p. 623.)

illustrate this fusion well with the subesophageal structure composed of six ganglia.

Neurosecretory cells have been found in the central nervous system to include subesophageal ganglia and ventral nerve cords of the three major annelid classes. Because of its apparent preeminence, hormonally as well as neurally, the cerebral ganglion has been implicated as the principal seat of control for the processes under neuroendocrine regulation. The three major physiological activities responsive to neurosecretion are regeneration, reproduction, and somatic transformation, or epitoky, and they will be discussed, where appropriate, under each class.

I. POLYCHAETA

This annelid class, the largest of the three, serves as a vital link in the marine ecosystem as primary or secondary consumer, and is preyed upon by almost all other marine invertebrates and fish. They can be divided into Errantia and Sedentaria, groupings based on adaptations for an active or sedentary life, and not on two discrete lines of evolution.

A. Morphology

The most distinctive morphological characteristics are the paired appendages on each somite, known as parapodia, with projecting setae, consisting of a rather distinct head containing several specialized appendages and one or two pairs of prostomial eyes. The head has a preoral projection, the prostomium, which contains the cerebral ganglia and gives rise to the ventral palps and dorsal tentacles (Fig. 8-4).

1. NEUROSECRETORY CELLS

The morphology of neurosecretory cells in seven polychaete families is discussed by Gabe (1966) in his monograph. After acknowledging the numerous reports over the previous three decades, primarily concerning the family Nereidae in which authors had classified neurosecretory cells into generally three or four types, Gabe discussed the difficulties faced in standardizing the nomenclature or synthesizing relationships existing among the cells de-

76   8. ANNELIDA

scribed. This difficulty was the very substance of an earlier presentation by Clark (1963) on the problem of evaluating evidence for neurosecretion in Nereidae as well as all annelids.

> There appear to be very large numbers of neurosecretory cells in the annelid brain, using the criterion that the cells contain granules stainable with paraldehydefuchsin, etc. It is doubtful if these are all neurosecretory cells. Of the four types of cell originally described in the brain of *Nereis*, only one now appears unequivocally neurosecretory. Making the criteria more stringent by insisting that the cells should undergo a secretory cycle is unhelpful because most of them can be shown to do so.

Since Gabe and Clark were so reluctant to construct a generalized scheme of polychaete cell types, equal caution is expressed in this work so as not to perpetuate a cell classification which may soon be altered or completely deleted from the annelid neuroendocrine literature. At the same time complete rejection of the morphology in an introductory text would be unnecessary. Thus, the following somewhat generalized observations will be made concerning principally *Nereis diversicolor* and *Platynereis dumerilii* of the family Nereidae.

Cells located in the posterior ventral and anterior dorsal borders of the cerebral ganglia appear to manifest many of the characteristics of neurosecretory cells (Fig. 8-2). They contain material which stains violet with Altmann's acid fuchsin, and black or reddish with chrome-hematoxylin-phloxine. Axons have been followed into the neuropile where some cross to the other half of the brain and then descend to terminate in the neurohemal area. A fluctuation in the amount of staining material has been noted, with the greatest concentration appearing in the immature and actively regenerating specimens when comparisons are made with the mature worm.

The difficulty not only occurs in trying to determine which cells are neurosecretory but in classifying those cells so designated into groups with stable morphological characteristics, having similar physiological properties. Golding (1967a) provides a discussion of this matter which is helpful and which will probably lead to an approach to annelids as valid as the approach to gastropods delineated by Simpson, Bern, and Nishioka (1966). He summarizes the 4-cell classification of B. Scharrer (1937) in the following sequence. The b and d cells should not be considered neurosecretory for lack of convincing evidence; in fact, the d cells are now regarded as photoreceptors. The a and c cells, found in nuclei XIXS and XXII, may be neurosecretory, but the final status is still considered uncertain.

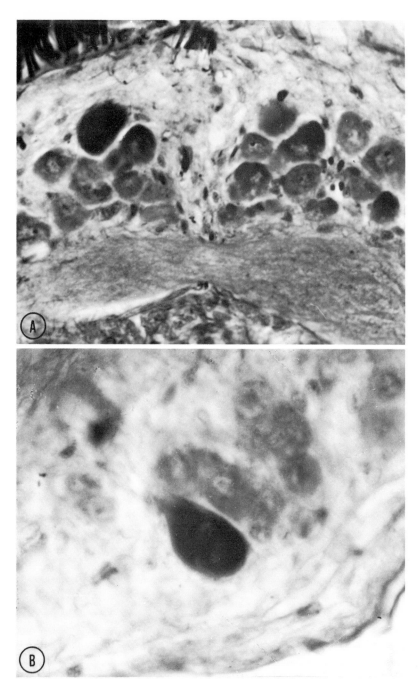

Fig. 8-2. Neurosecretory cells in the brain of *Platynereis dumerilii*. A. Anterior region, × 1000. B. Posterior region, X 940. Altmann's acid fuchsin, methyl green. (Hauenschild and Fisher, 1962.)

Similarly Dhainaut-Courtois (1966) identifies six types of brain cells in *Nereis pelagica*, but considers only one to be neurosecretory.

2. NEUROHEMAL ORGAN

There are many indications that a primitive neurohemal structure is present in the polychaetes in the form of a cerebrovascular, or infracerebral complex. (See Figs. 8-3 and 8-18 for a similar structure in a leech.) Such a structure has been described in the ventral part of the cerebral ganglion of two polychaetes in proximity to the posteriorly located neurosecretory cells, thus providing these cells with a possible storage and release complex in association with the vascular system. Secretions have been most frequently detected in the cerebrovascular complex of immature polychaetes, but such evidence is not present for the mature worms. This supports the previously mentioned evidence that the cells of immature worms are the most active and probably use this primitive neurohemal structure to facilitate transport of the juvenile hormone throughout the body.

There is morphological evidence available suggesting the presence of intrinsic neurosecretory cells within the infracerebral gland of several nereids. Such fuchsinophilic cells may be responsive to the neurosecretion which originates within the brain and passes ventrally into this area. A second type of cell is present which definitely is not neurosecretory, has all the characteristics of epithelial cells, and appears suggestive of nonneural endocrine tissue responsive to and possibly associated with the described neurosecretory elements (Golding et al., 1968). This arrangement is similar to the brain-corpus cardiacum-corpus allatum complex of insects.

B. Physiology

Three physiological functions in polychaetes have been shown to be under neuroendocrine control of the cerebral ganglion: epitoky, reproduction, and regeneration. All three are associated with one another in that they are directly related to the propagation of the species.

1. EPITOKY

Polychaetes live most of the year as sexually immature sedentary animals known as atoke, near the bottom of bodies of water. When

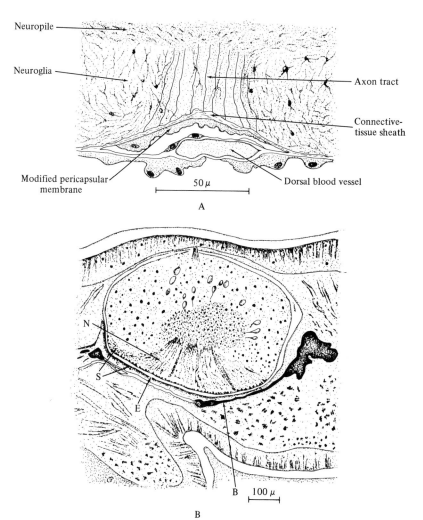

Neuropile

Neuroglia

Axon tract

Connective-tissue sheath

Modified pericapsular membrane

50 μ

Dorsal blood vessel

A

B 100 μ

B

Fig. 8-3. The cerebrovascular complex of two polychaetes: A, *Nephtys californiensis* (ventral part of the supraesophageal ganglion in cross section). (From Clark, 1966.) B, *Perinereis cultrifera* (sagittal section of the supraesophageal ganglion, and part of the prostomium). N, Fiber tract; S, granules of secretion; E, modified pericapsular membrane; B, dorsal blood vessel. (From Bobin and Durchon, 1952.)

breeding season approaches, not only must the gonads develop, but the worms must be morphologically modified so that they are able to swim to the surface and spawn. This metamorphosis of the immature worm into the reproductive form is known as epitoky or epigamy, an older and less used term.

The advantages that such a spawning migration provides to the animal are obvious. Since they are dioecious and internal fertilization is rare, the products of their reproductive systems are shed to the sea and fertilization is random. If such reproductive behavior is to be successful, then synchrony must be present in the formation of the epitoke, the reproductive form, and the arrival of the epitoke in a swarm to the spawning area.

Although epitoky as described below is principally a characteristic of the Nereidae, it occurs in all families of Polychaeta with some modifications. The external transformation produces a larger body, enlarged eyes, a change in the parapodia, and a reduction in the cephalic appendages (Fig. 8-4A and B). Internal morphological changes occur in the vascular, muscular, and digestive systems, with maturation of the gonads. All of these modifications produce a freely swimming, pelagic animal capable of rising to the surface, swimming excitedly, and then shedding its gametes to the open sea.

As to the control of epitoky, it has been shown a number of times, by extirpation and transplantation of the cerebral ganglion among various aged worms, that a hormone which influences metamorphosis is secreted by the brain. It is a hormone which exerts not only an inhibitory influence on the growth and development of the worm but is also necessary as a positive factor for the proper implementation of metamorphosis after total inhibition has been removed.

In *Platynereis dumerilii* a high level of the hormone inhibits epitoky, but a lower level must be retained to initiate the normal chain of events leading to complete metamorphosis. If, during metamorphosis, the level of hormone is increased experimentally, then the morphological transformation is arrested. However, there is a point in the process beyond which the transformation can not be influenced by reimplanting immature cerebral ganglia.

2. GONADAL MATURATION

When epitoky is in progress, the gonads of certain polychaetes mature with the subsequent formation of vast numbers of spermatozoids and oocytes. These are stored in the coelom until the correct body signal is received for their massive release into the open sea for fertilization.

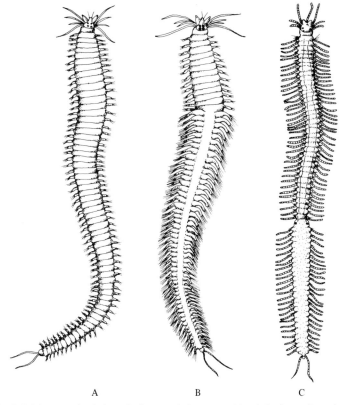

Fig. 8-4. Metamorphosis in polychaetes. A. Immature *Nereis* in the atokous form. B. Mature *Nereis* in the epitokous form. C. Stolonization in syllid polychaetes. (From Charniaux-Cotton and Kleinholz, 1964; after Durchon, 1960.)

This is the most common but not the sole reproductive process. In other worms a surer but slower way of producing young is through the development of a budding zone midway in the body with the transformation of the posterior somites into an epitoke which then breaks away from the parent with the formation of its new head. This process is repeated and modifications are found among the polychaetes (Fig. 8.4C).

a. Spermatogenesis

The initial experimental evidence indicating possible endocrine control over reproduction was obtained when precocious genital maturation followed the removal of male germ cells of *Nereis* from the influence of the cerebral ganglia by cutting the animal in half, by

surgically removing the cerebral ganglia, or by ablation of the prostomium. When the male animals were cut in half and left for several
weeks, the anterior portion with the cerebral ganglion contained only
spermatocytes, while the posterior half produced mature spermata-
zoids comparable to those of a swarming adult. When cerebral
ganglia of immature worms were implanted into males approaching
sexual maturity, the maturation of spermatocytes was inhibited;
likewise, when such ganglia were implanted into decerebrate animals,
precocious maturation of spermatocytes was prevented. All of these
observations indicate that a direct effect of the cerebral hormone is
the retarding of male sex cell maturation.

b. Oogenesis

In general this system is very similar to that in the male, with a high
level of hormone in the immature worm inhibiting development of the
oocytes. This hormone has its origin in the cerebral ganglion and, as
its concentration diminishes, epitoky is initiated, coincidental with an
increase in gonadal development. A low level of the hormone remains
essential for successful vitellogenesis, although the final stages of
maturation are not completely dependent upon its presence (Fig. 8-5).
In *Nereis diversicolor,* after the first 100 days when the oocytes are
less responsive to the inhibiting hormone, the growth curve becomes

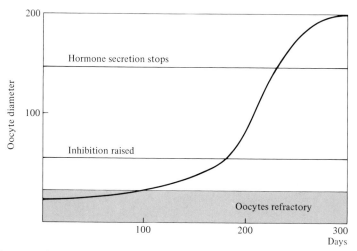

Fig. 8-5. Scheme of the hormonal control of oogenesis in *Nereis diversicolor.* With
suitable adjustment of the scales on the abscissa and ordinate, the same diagram is
probably applicable to most nereids. (From Clark, 1965.)

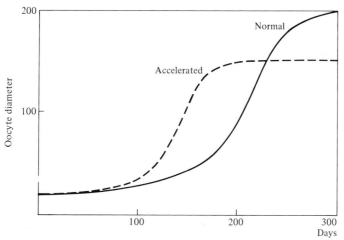

Fig. 8-6. Normal and accelerated growth of oocytes in intact and decerebrate *Nereis diversicolor,* respectively. (From Clark, 1965.)

sigmoid, characterizing the period of vitellogenesis. As the hormone concentration is further reduced, maximum sized oocytes are attained at approximately 300 days.

Accelerated growth of oocytes can be obtained once their diameter has reached 30 μ by extirpation of the cerebral ganglion. The eggs resulting from such a procedure are deficient in protein and are considerably reduced in size. Nevertheless, fertilization can occur with initial cleavages, but development usually fails around the 8- or 16-cell stage. If the oocytes have reached a size of 160–170 μ when ganglion extirpation occurs, the eggs produced are normal in all respects (Fig. 8-6).

Thus both epitoky and sexual maturation are inhibited by a high level of hormone. Once this concentration begins to decline, maturation processes are initiated, but a low level must be maintained to provide the proper hormonal environment for both processes to reach their successful conclusions. Only during the final stages of both epitoky and gametogenesis can they continue to a satisfactory conclusion without any portion of the cerebral ganglion.

With both processes under similar hormonal control, it has been assumed, on the basis of indirect evidence, that the same chemical entity regulates both gametogenesis and epitoky. Preliminary attempts to fractionate the extracts of the supraesophageal ganglia have not been successful.

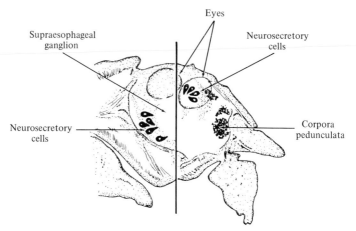

Fig. 8-7. Parasagittal section of the prostomium and supraesophageal ganglion of *Platynereis dumerilii*, showing the two major groups of neurosecretory cells which are probably the source of juvenile hormone. If the ganglion is transected along the line indicated, both halves show juvenile hormone activity. (From Clark, 1965, after Hauenschild, 1964.)

Attempts to link the origin of this hormone(s) with a neurosecretory cell type have been unsuccessful also. However, when the cerebral ganglion in *Platynereis dumerilii* is sectioned so that the posterior and anterior neurosecretory cells are separated, juvenile hormone activity is retained by both halves when implanted into decerebrate worms (Fig. 8-7). Thus, the hormone must be present at more than one site within the cerebral ganglion although the cell type giving rise to this hormone is not definable. As to the mechanism of action of this inhibiting hormone, evidence available from histochemical, biochemical, and ultrastructural technique, in conjunction with organ culture, indicates that the hormone acts by regulating the synthesis of ribonucleoprotein. In the absence of the hormone, distinct changes are apparent with an increased amount of RNA around the oocyte nucleus which is not evident when the oocyte has been in the presence of the hormone (Durchon, 1969).

c. Environmental Factors

The amazing regularity and synchrony found in the reproductive biology of polychaetes offer the student of neural integration some of the most challenging problems in all of biology. A beginning has been made in analyzing the effects of environmental factors on spawning, and a most interesting and enlightening discussion of this is found in the review by Clark (1965).

It was concluded in the last section that in nereids and other Polychaeta, metamorphosis and maturation generally begin when a decline occurs in the secretions of the juvenile hormone. It follows, therefore, that the activity of the cerebral ganglion is very probably affected by one or a combination of environmental stimuli such as length and intensity of lunar or sun photoperiods, temperature levels, and availability of food. Just how the interpretation of these environmental signals is made represents a problem which is not unique to Annelida but transcends all animal phyla. The time at which this interpretation occurs is also open to considerable discussion, but two possibilities are illustrated in Fig. 8-8. If epitoky, vitellogenesis, and spawning follow in sequence after an environmental signal initiates the cerebral ganglion's gradual inactivation, then a poorly synchronized period of breeding is observed. However, closer coordination may exist, as in *Odontosyllis enopla,* where swarming begins in mature worms slightly less than an hour following sunset, and spawning is completed within 20 minutes. The possibility then exists that the worms await the reception of a final environmental signal which occurs after epitoky and vitellogenesis have been completed. Whether this final reproductive act is under hormonal or nervous control is not known.

The most logical transducer for these photoperiodic stimuli are the photoreceptive cells of the eyes, although this is not necessarily the case with *Platynereis dumerilii.* Under a fluctuating light regime in the laboratory the same endogenous cycles are imprinted in blinded as in normal worms; this suggests that light may be directly influencing the neurosecretory cells (Hauenschild, 1964).

3. REGENERATION

During early growth of the marine polychaetes, when segments are being added at regular intervals and the organisms are under the influence of the previously discussed inhibitory factor, polychaetes also possess considerable regenerative abilities. It has been known for some 20 years that through the removal of the prostomium, the process of regeneration following amputation of posterior segments could be inhibited; this condition could then be reversed by the reimplantation of the prostomium or, essentially, the cerebral ganglia. This was a clear indication of hormonal involvement, and this observation has

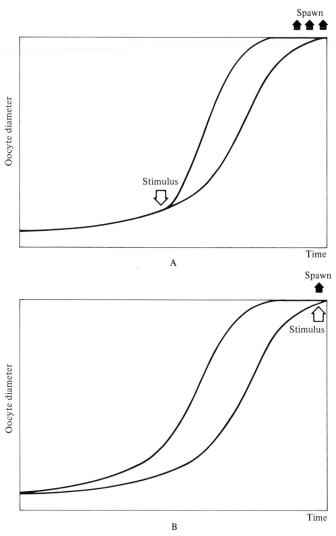

Fig. 8-8. Influence of the stage at which environmental stimuli impinge upon the maturation processes on the synchrony of spawning. A. The environmental stimulus (open arrow) initiates maturation in all members of the population synchronously. Variation in the rate of maturation desynchronizes the population and spawning (closed arrows) occurs over an extended period. B. The environmental stimulus initiates spawning. Although previously unsynchronized, all members of the population that are mature at this time spawn synchronously. (From Clark, 1965.)

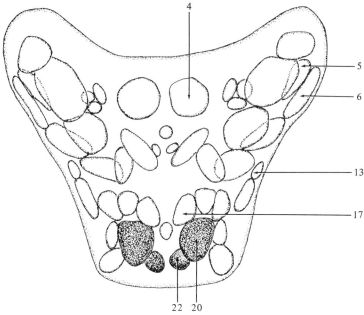

Fig. 8-9. Plan of the supraesophageal gangloin of *Nereis diversicolor,* showing ganglionic nuclei 20 and 22 (stippled) in which neurosecretory cells have been reported to show a response after the amputation of posterior segments. Nuclei 4, 5, 6, 13, and 17 are now considered uncertain to contain neurosecretory cells. (After Clark and Bonney, 1960; with subsequent addition from Clark, 1966, Dhainaut-Courtois, 1966, and Golding, 1967a, b, c, d.)

since been confirmed by a large body of literature. Subsequent investigations have followed two approaches: first, to describe the changes in neurosecretory cells during regeneration, and second, to explore the course of the hormonally controlled regeneration.

The description and location of the cells of the cerebral ganglia which are active in regeneration constitute a problem of terminology and cell identification. However, a number of cells within the cerebral ganglion nuclei clearly show cyclical patterns in synthesis and release, beginning as early as 6 hours following amputation; they are believed related directly or indirectly to the phenomenon of regeneration. Their location can best be summarized in Fig. 8-9, a diagram of the nuclei in the cerebral ganglia of *Nereis diversicolor.*

A considerable body of information is available concerning the course of events leading to the neuroendocrine response to amputation. However, the entire picture has recently been reexamined in detail by Golding (1967b,c,d), and many of the earlier interpretations have now been revised with the production of a rather simple first-order neuroendocrine reflex. The most prevalent concept in the literature which now must be altered is the idea that the hormonal content of the cerebral ganglion gradually increases from the day of amputation to about the fourth day, when a considerable portion of the hormone is released into the general circulation. Also, the idea that the released hormones serve to "trigger" a "critical" stage in the regeneration process, possible at the developing blastema, now seems very unlikely. Clark (1966), in a recent review, has very ably interpreted the new evidence by Golding in light of the previous literature, and much of the following is taken from his article.

Apparently a general growth hormone exists which is continually secreted by the cerebral ganglia at a consistent rate during the life of the immature worm. This gives rise to the regular pattern of body growth by proliferation of new segments. Whenever segments are lost accidentally or experimentally there is a change in the sensitivity of the tissue at the wound to the growth hormone, with a subsequent

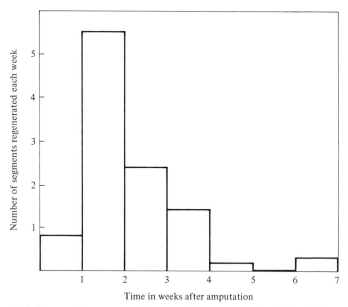

Fig. 8-10. The rate of segment proliferation during regeneration in Nereis diversicolor with in situ cerebral ganglia. (From Golding, 1967c.)

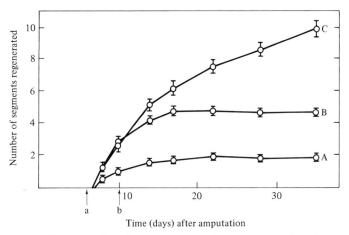

Time (days) after amputation

Fig. 8-11. Effect upon the subsequent course of regeneration when the cerebral ganglion of *Nereis diversicolor* is extirpated at various times after amputating posterior segments. A. Ganglion extirpated on the sixth day; B. Ganglion extirpated on the tenth day after amputating segments; C. Ganglion intact throughout. (From Clark, 1966, after Golding, 1967b.)

change in the rate of segment proliferation. Conversely, when the number of segments begins to reach the normal level, the rate of regeneration declines as the sensitivity of the tissue to the hormone is reduced (Fig. 8-10). When the cerebral ganglion is removed from the body of a juvenile worm which has undergone posterior segment amputation, the ability of the worm to continue regeneration is altered (Fig. 8-11). Following the reimplantation of the ganglion, this deficiency is corrected.

Therefore, when the juvenile or the growth-inhibiting hormone begins to decline in concentration, with the subsequent onset of epitoky or reproduction, a decline will also be apparent in the growth hormone, or as Golding has termed it, the "regeneration hormone." All energy available to the organism will then be diverted from the growth and regenerative process to the manifold metamorphose and reproductive activities.

II. OLIGOCHAETA

The second and most familiar annelid class contains the terrestrial earthworms which live in moist soil almost everywhere but are

especially abundant in temperate regions. Some are aquatic and rather
widespread; a few have even borrowed a trait from the leeches and
have become parasitic on aquatic animals.

A. Morphology

The Oligochaeta morphology is quite simple with no parapodia
or prostomial eyes, as were common on polychaetes. The head has
been reduced and the prostomium is often hardly more than a tiny
flap. As shown earlier in Fig. 8-1, the cerebral ganglion has moved
posteriorly with numerous sensory fibers innervating the anterior
segments.

1. NEUROSECRETORY CELLS

Four families have been shown to possess neurosecretory cells.
The Naididae are somewhat novel in that they contain cells, staining
with hypodermal cells which also show an affinity to the above stain
than in the cerebral ganglion. These cells are quite small, with an
average diameter of 6 μ. They possess short axons extending distally
from the nerve cord toward the hypodermis, coming to rest in contact
with hypodermal cells which also show an affinity to the above stain
(Deuse-Zimmermann, 1960). Of the remaining three families, Lumbricidae has received by far the most attention with *Lumbricus terrestris*
as the most thoroughly studied species.

Despite the diversity of staining techniques and schemes for cell
identification, there is less difficulty in classifying these cells than
there is with Polychaeta. Bern and Hagadorn (1965) utilized all of the
available evidence and developed a useful table showing the possible
equivalence of the different cell types which have been proposed by
various authors (Table 8-1). Bern and Hagadorn, as well as Gabe
(1966), feel that the Herlant-Meewis classification should be the guide
for future attempts at cell study of oligochaetes. She relies on well-
defined cell features, minimizes variation in staining affinity, and
supports her classification with histophysiological observations. Three
neurosecretory cell types are thus associated with the cerebral ganglion: a-cell, b-cell, and the medium and large neuron. One other
type, the u-cell, is in the subesophageal ganglion.

The a-cells are situated in the posterior part of the cerebral ganglion
and show differences in staining affinities which are indicative of a
secretory cycle. This grouping of cells has been referred to by several

TABLE 8-1
Possible Equivalence of Neurosecretory Cell Types in Oligochaeta[a,b]

Herlant-Meewis	Hubl	Harms	Brandenburg	Michon and Alaphilippe	Aros and Vigh	Description	Physiological correlates
a-cells	a-cells (in part) b-cells (in part)	a-cells (in part) Blue cells (in part)		a-cells		Located in posterolateral portion of cerebral ganglion; dense cytoplasm; nucleus often eccentric; axons do not leave supraesophageal ganglion	Reproductive cycle (Herlant-Meewis, Hubl)
b-cells	a-cells (in part) b-cells (in part) c-cells	a-cells (in part) Blue cells (in part)		b-cells		Located laterally below a-cell region near point of emergence of anterior connectives; fusiform; radial orientation. Some axons leave supraesophageal ganglia via circumesophageal connectives	Regeneration (Hubl)
Subesophageal neurosecretory cells	u-cells (?)	Red and blue cells	Stage 6 = u-cells of Hubl	d-cells (?) (seen in one species only)		Located in subesophageal ganglion	Regeneration (Hubl)
Large and medium-sized neurons	Large and small neurons			Δ-cells	Dorsomedial cell group	Internal to and larger than a- and b-cells	Color change (Aros and Vigh); egg-laying (Herlant-Meewis, Hubl)

[a] From Bern and Hagadorn, 1965.
[b] With special reference to the Lumbricidae.

as a "cerebral organ," but it is doubtful that this area warrants such a designation. Ventral and lateral to the a-cells, and closer to the origin of the periesophageal trunk, type b-cells, usually fusiform in shape, are found. Location and morphology are the principal differentiating qualities of these two cell types. More internally or medially located within the cerebral ganglion is the third group of cells, referred to as the medium- and large-sized neurons. Only one class of cells is set up for the subesophageal ganglion, although a number of terms and subclasses have been suggested by various authors.

The initial ultrastructural studies on the annelids were accomplished by E. Scharrer and Brown (1961) on *Lumbricus terrestris,* and their findings have been alluded to on several occasions in this text. Essentially, their observations indicated that the membrane-bounded secretory granules were formed by the Golgi apparatus after synthesis of the components elsewhere (Fig. 2-1). Their observations as well as subsequent ultrastructural studies have unfortunately added little to the knowledge of cell classification.

2. NEUROHEMAL AREA

The presence of a neurohemal organ in the Oligochaeta has not been demonstrated as it has been in the Polychaeta; however, the cerebral ganglion is intensely vascularized, internally as well as externally. This stems from the ventral blood vessel which forms a plexus behind the ganglion, giving rise to the network. It is reasonable to assume that distribution of neurosecretion is facilitated by this close association with the circulatory system.

B. Physiology

1. REPRODUCTION

Oligochaetes exhibit an annual reproductive cycle with the gonads maturing in spring or early summer, regressing later in the year, and maturing again the following spring. A cycle of neurosecretory cell development and activity has been shown by Herlant-Meewis (1957, 1962, 1964) to begin with postembryonic development and extend through oviposition.

In the young of *Eisenia foetida,* with the body less than 20 mm long, and with only primitive gonocytes in the gonads, a-cells within the cerebral ganglion are small but easily identifiable. The b-cells are not yet completely differentiated, and identification is difficult. At 50 mm in length, the differentiation of the genital tract is complete and vitellogenesis is underway. Both a- and b-cells are now distinct; a-cell axons can be followed through the neuropile, while the b-cells have appeared in their fusiform shape. When full sexual maturity has been reached, the body is between 80–90 mm in length, and the necessary secondary sexual structures, i.e., clitellum, are fully functional. At this time the a-cells also exhibit their definitive characteristics (Fig. 8-12).

With such a parallel existing between the genital and neurosecretory cell development, a relationship between these two systems is suggested. When morphological observations are continued on this species, copulation and oviposition coincide with vacuolation of the perikaryon, resulting in a reduction of staining in the cell body as well as axons. And at the end of the period of oviposition, when the egg chamber has been exhausted, very little neurosecretion remains in the a-cells.

Further support for the contention that a hormone does regulate reproductive development is provided by experimental procedures. Removal of the cerebral ganglion immediately arrests egg laying, with a loss of weight and secondary sex characteristics, such as the clitellum and tubercula pubertatis (Fig. 8-13). In 8 weeks when the cerebral ganglion has been regenerated, egg laying begins again. On closer examination, the regeneration of the cerebral ganglion is very advanced at 3 weeks, but the periganglionic capillary network has not penetrated into the neurosecretory cell area. It is not until this intraganglionic capillary network has been fully reconstructed in the area of the a-cells that ovulation is reinstated. When larger portions of the central nervous system are removed, longer periods of reproduction arrest occur.

It therefore seems evident that the cerebral ganglion is instrumental, if not indispensable, in the regulation of the reproductive cycle of the earthworm. The a-cells and not the b-cells appear to be the active components, and the neurosecretory substance is probably transported from the site of release to the target organs by the vascular system.

2. DIAPAUSE

A number of earthworms enter a period of reproductive inactivity in summer which extends through the winter and is suspected to be a response to the environmental conditions of the soil. This is often referred to as a period of aestivation or, more appropriately, as

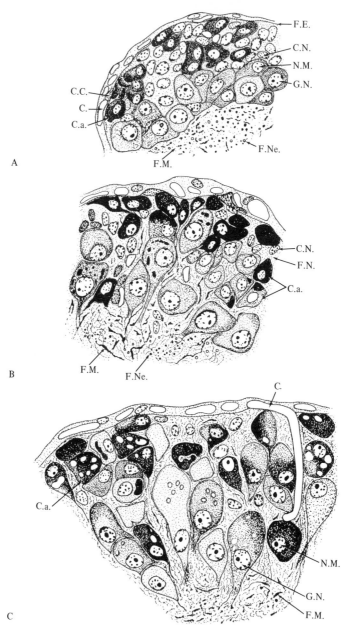

Fig. 8-12. Cyclic changes of cerebral neurosecretory perikarya during growth of *Eisenia foetida*. A. Freshly hatched specimen; B. 47-mm body length; C. specimen fixed after laying of 47 cocoons. C.a., a-cells; C., capillaries; C.C., capsular cells; C.N., glial cells; G.N., large neurons; N.M., medium-sized neurons; F.M., beaded fibers; F.Ne., nerve fibers; F.E., elastic capsular fibers. (From Herlant-Meewis, 1956.)

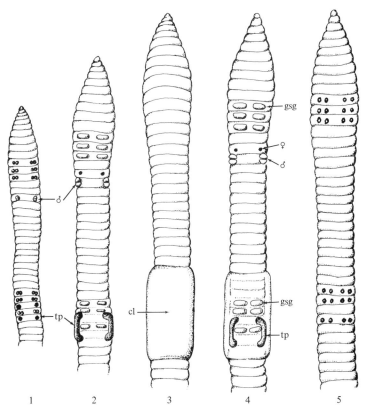

Fig. 8-13. Development of somatic sexual characteristics on *Allolobophora terrestris typica*. 1–2, Young worms with the development of the ventral cutaneous glands; 3, adult in reproductive activity (dorsal view); 4, adult in reproductive activity (ventral view). 5, adult in reproductive rest (ventral view); cl, clitellum; gsg, genital glands; tp, tubercula pubertatis. (From Durchon, 1967.)

aestivo-hibernation. Since a certain time period must elapse before activity is returned, the suggestion has been made that a hormone is involved in the regulation of this annual period of inactivity. After examination of the cerebral ganglion of the diapausing worm, it was noted that the a-cells were rich in neurosecretion. This accumulation apparently occurred between the end of summer ovoposition and diapause, but release did not come until the termination of diapause (Fig. 8-14). And, again in support of the previously stated role of the a-cells in reproduction, when the material is transported from the

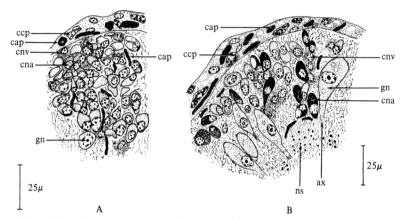

Fig. 8-14. A. Sketch of a parasagittal section of the posterior region of the brain of *Allolobophora icterica* during diapause. The neurosecretory cells of type a are of reduced size with little evidence of secretion. B. Section of the posterior region of the brain of the same species, undergoing voluntary diapause, at the time of the blastomene appearance during caudal regeneration. The neurosecretory cells of type a are either vaculated or show little evidence of secretion. ccp, Cells of the cerebral capsule; cna, neurosecretory cells of type a; cnv, neuroglial cells; gn, large nerve; cap, capillaire; ns, neurosecretion: ax. axon. (From Durchon, 1967.)

ganglion, the reproductive cycle is once more initiated. Whether it is the release of the a-cell content which breaks diapause as well as initiating reproductive development, or if release of the a-cell secretion is a result of a prior hormonal activation is as yet not known.

3. REGENERATION

A problem which has attracted the attention of experimentalists for over half a century is regeneration in annelids and the earthworm. It now appears that in *Lumbricus* the anterior portion of the nervous system is necessary for regeneration. The b-cells of the cerebral ganglion are the key cells and serve as the origin of a hormonal factor which is released during the first 2 days following posterior amputation. The series of experiments which Hubl (1956) performed in establishing this hormonal relationship were summarized as follows by Gabe (1966):

 1. Amputation of the anterior extremity and the last segments of the body. — Eight weeks after the operation regeneration of the anterior part was advanced but regeneration of the posterior had not started.
 2. Removal of the cerebral ganglion and amputation of the posterior part of the

body. — Eight weeks after the operation regeneration of the brain was well advanced but regeneration of the posterior part had not started.

3. Amputation of the posterior part of the body and amputation of the anterior part 48 hours later. — Eight weeks after the operation both anterior and posterior regeneration were well advanced.

4. Amputation of the posterior part of the body with removal of the cerebral ganglion 48 hours later. — The brain regenerated normally and posterior regeneration proceeded as if only the posterior extremity of the worm had been injured.

Thus, 48 hours are sufficient for the proper cells to be informed of the loss of segments, to synthesize if necessary the required hormone, and to release a sufficient amount to allow the organism to begin the process of posterior regeneration. After this 2-day period the brain can also be removed with subsequent cerebral regeneration.

4. COLOR CHANGE

Following the exposure of Lumbricus rubellus to the sun or ultraviolet light, the dorsal medial group of neurosecretory cells of the brain shows vacuolation, and the worms turn dark. After the animals are again placed in the dark they turn pale and the cells are filled once more with secretion. No studies have been performed as yet on operated animals to support the histological evidence.

However, the possibility that the above is simple chromatophorotropic effect suggests that this relationship may not be justified. Neurosecretion could be affecting some other system, and the change in coloration may be due to photosensitization of a dynamic biochemical (Laverack, 1963).

5. OSMOREGULATION

The influence of the brain over water balance has been shown to exist in several of the previous phyla, and a similar suggestion has recently been offered for Lumbricus terrestris. By removing various parts of the central nervous system, Kamemoto (1964) showed that when the cerebral ganglion was extricated from a worm which was then placed in tap water a marked increase in body weight was observed over that of normal animals under similar conditions. When the subesophageal ganglion was removed, or when the circumesophageal connectives were severed, the weight change was comparable to that of the normal animal (Fig. 8-15). Increase in weight of a brainless worm can be reversed by the implantation of a brain into the coelom, or by the injection of brain homogenates into the body. However,

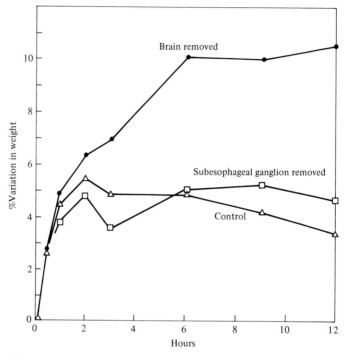

Fig. 8-15. Variations in weight of *Lumbricus terrestris* placed in distilled water, showing evidence of the role of the brain in the regulation of osmotic pressure in the earthworm. (From Durchon, 1967, after Kamemoto, 1964.)

histological examinations show no variation in the neurosecretory cells between the implanted and normal brains.

III. HIRUDINEA

As the smallest class of annelids with about 300 species, the leeches are predaceous animals from predominantly freshwater habitats. History has popularized the medical use of the blood-sucking leech, but many are ordinary predators or scavengers. Of those that are considered predaceous, many take blood meals from invertebrates as well as vertebrates.

A. Morphology

The body of a leech is usually flattened dorsoventrally and tapered at each end with the body length between 20 and 60 mm. The inward movement of the central nervous system and centralization, two evolutionary trends in annelids, are quite evident in the leeches (Fig. 8-1). Movement of the cerebral ganglion posteriorally, and the union of the anterior ganglia of the ventral chain with the subesophageal ganglion, reflects, very probably, the extensive development of the oral sucker. This explains the highly compartmentalized brain: 12 compartments for the cerebral ganglion and 24 for the subesophageal ganglion (Fig. 8-16).

1. NEUROSECRETORY CELLS

Of the dozen species studied in respect to neuroendocrinology, the most thoroughly studied are *Theromyzon rude* and *Hirudo medicinalis* (Hagadorn, 1962, 1966; Hagadorn et al., 1963). There

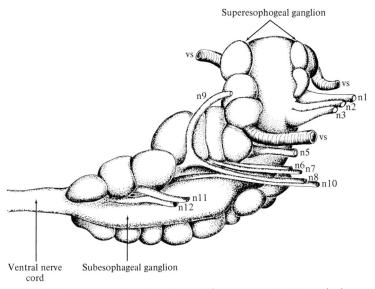

Fig. 8-16. Schematic drawing of the brain of *Theromyzon*. n1–n12, cerebral nerves; vs, blood vessel. (From Durchon, 1967, after Hagadorn, 1958.)

appear to be two major types of neurosecretory cells, α and β (Fig. 8-17). The α cells are found in the cerebral ganglion where the maximum diameter is about 35 μ, and in the subesophageal ganglion with a maximum diameter of 15 μ. Both are ovoid cells with oval nuclei centrally or eccentrically located with one or two nucleoli. The cell and transporting axons stain markedly with paraldehyde-fuchsin, indicating a secretion with a high cystine content. The fibers can therefore be easily followed from the brain, along the ventral nerve cord and into some segmental nerves.

The β cells are acidophilic; their secretion is poor in cystine and it is principally on that basis that the two types of cells are differentiated, for they do not differ significantly in size or location. Two subclasses have been identified: β_1 which stains with orange G and is poor in tryptophan, and β_2 which stains with light green SF and is rich in tryptophan. Whether these subclasses are two morphologically and biochemically distinct cell types or two stages in the functional cycle of β cells is not known.

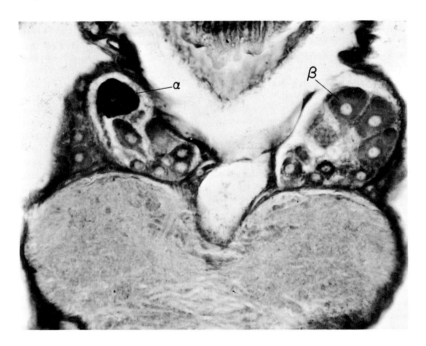

Fig. 8-17. Frontal section of the supraesophageal ganglion of *Theromyzon* to show α and β cells. DCA, dorsal commissure. Anterior toward bottom. (From Hagadorn, 1966.)

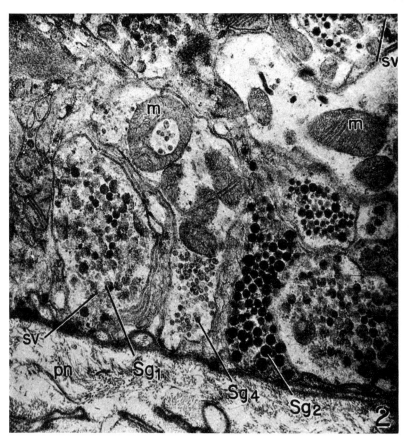

Fig. 8-18. Neurohemal area of the dorsal commissure in *Theromyzon*. Axons are shown ending under the perineurium. Three types of secretion are shown, each segregated in a specific axon ending. Sg_1, Large vesicular secretion; Sg_2, large granular secretion; Sg_4, small vesicular secretion; m, mitochondrion; pn, perineurium; sv, presumed synaptic vesicles. (From Hagadorn, 1962.)

Electron microscopic examination of the cerebral ganglion indicates a vast number of elementary neurosecretory granules with an average diameter between 2000 and 3000 Å. There are at least four distinct types when classified on the basis of maximum diameter, presence or absence of a limiting membrane, and degree of electron density. Three of the types are shown in Fig. 8-18, which also illustrates the suspected neurohemal area. Normal organelles for such cells are present with mitochondria appearing large and cup shaped.

2. NEUROHEMAL AREA

In *Theromyzon* a loose network of secretion-bearing axons appear to terminate on the posterodorsal surface of the dorsal commissure in an area of close association with a transverse vessel (Fig. 8-18). This indicates that a primitive neurohemal structure is present which may by quite similar to structures in Fig. 8-4 for two polychaetes.

B. Physiology

1. REPRODUCTION

When examining the α cells of *Theromyzon* during the year, a definite cycle is observed. The maximum number of neurosecretory cells is observed from April through June which precedes, by a month, the advent of reproductive activity. Regression begins in midsummer and reaches a low level in October, remaining throughout the winter until once more a burst of activity is observed in April. This cyclical variation coincides closely with the annual reproductive cycle and, indeed, is quite similar to the cycle observed in the earthworm. In addition, because of evidence from an experimental study on the influence of the brain on spermatogenesis, Hagadorn has concluded that a gonadotropic influence does originate in the brain of *Hirudo* as well as *Theromyzon,* and that the α cells are the suggested source of this factor (Fig. 8-19).

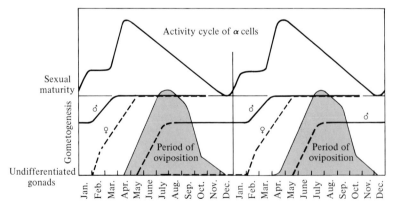

Fig. 8-19. Cycle of activity of the α neurosecretory cells and the reproductive cycle of *Theromyzon.* (From Durchon, 1967, after Hagadorn, 1962.)

References

Bern, H. A., and Hagadorn, I. R. (1965). Neurosecretion. In "Structure and Function in the Nervous System of Invertebrates" (T. H. Bullock and G. A. Horridge, eds.), Vol. 1, pp. 353–429. Freeman, San Francisco, California.

Bobin, G., and Durchon, M. (1952). Etude histologique du cerveau de Perinereis cultrifera. Mise en évidence d'une complèxe cérébro-vasculaire. Archs. Anat. microsc. Morph. exp. 41, 25–40.

Charniaux-Cotton, H., and Kleinholz, L. H. (1964.) Hormones in invertebrates other than insects. In "The Hormones" (G. Pincus, K. V. Thimann, and E. B. Astwood, eds.), Vol. 4, pp. 135–198. Academic Press, New York.

Clark, R. B. (1963). Problems of interpreting neurosecretory phenomena in annelids. J. Endocrinol. 26, XVIII–XIX.

Clark, R. B. (1965). Endocrinology and the reproductive biology of the Polychaetes. Oceanog. Marine Biol. Ann. Rev. 3, 211–255.

Clark, R. B. (1966). The integrative action of a worm's brain. Symp. Soc. Exptl. Biol. 20, 345–379.

Clark, R. B., and Bonney, D. G. (1960). Influence of the supra-esophageal ganglion on posterior regeneration in Nereis diversicolor. J. Embryol. Exp. Morph., 8, 112–118.

Deuse-Zimmermann, R. (1960). Cited by Gabe (1966).

Dhainaut-Courtois N. (1966). Etude histologique des cellules nerveuses du cerveau de Nereis pelagia L. (Annélide, Polychete). Compt. Rend. 263, 1596–1599.

Durchon, M. (1960). L'endocrinologie chez les annelides polychetes. Bull. Soc. Zool. Fr., 85, 275–301.

Durchon, M. (1967). L'endocrinologie des Vers et des Mollusques. Masson, Paris.

Durchon, M. (1969). Endocrines and pharmacology of Annelida, Echiuroidea, Sipunculoidea. Chem. Zool. 4, 443–466.

Gabe, M. (1966). "Neurosecretion." Pergamon Press, Oxford.

Golding, D. W. (1967a). The diversity of secretory neurones in the brain of Nereis. Z. Zellforsch. Mikroskop. Anat. 82, 321–344.

Golding, D. W. (1967b). Neurosecretion and regeneration in Nereis. I. Regeneration and the role of the supraesophageal ganglion. Gen. Comp. Endocrinol. 8, 348–355.

Golding, D. W. (1967c). Neurosecretion and regeneration in Nereis. II. The prolonged secretory activity of the supraesophageal ganglion. Gen. Comp. Endocrinol. 8, 356–367.

Golding, D. W. (1967d). Regeneration and growth control in Nereis. J. Embryol. Exptl. Morphol. 18, 67–90.

Golding, D. W., Baskin, D. G., and Bern, H. A. (1968). The infracebral gland—a possible neuroendocrine complex in Nereis. J. Morphol. 124, 187–216.

Hagadorn, I. R. (1958). Neurosecretion and the brain of the rhynchobdellid leech, Theromyzon rude. J. Morph., 102, 55–90.

Hagadorn, I. R. (1962). Neurosecretory phenomena in the leech, Theromyzon rude. In "Neurosecretion" (H. Heller and R. B. Clark, eds.), pp. 313–321. Academic Press, New York.

Hagadorn, I. R. (1966). Neurosecretion in the Hirudinea and its possible role in reproduction. Am. Zoologist 6, 251–261.

Hagadorn, I. R., Bern, H. A., and Nishioka, R. S. (1963). The fine structure of the supraesophageal ganglion of the rhynchobdellid leech, Theromyzon rude, with special reference to neurosecretion. Z. Zellforsch. Mikroskop. Anat. 58, 714–758.

Hausenchild, C. (1964). Cited by Clark (1965).

Hauenschild, C. and Fisher, A. (1962). Neurosecretory control of development in *Platynereis dumerilii*. In "Neurosecretion" (H. Heller and R. B. Clark, eds.), pp. 297–312. Academic Press, New York.

Herlant-Meewis, H. (1956). Croissance et neurosécrétion chez *Eisenia foetida* (Sav.). *Ann. Sci. Nat. Zool. Biol. Animale* [11]**18,** 185–198.

Herlant-Meewis, H. (1957). Reproduction et neurosécrétion chez *Eisenia foetida. Ann. Soc. Zool. Belg.* **87,** 151–183.

Herlant-Meewis, H. (1962). Neurosecretory phenomena during regeneration of nervous centers in *Eisenia foetida. In* "Neurosecretion" (H. Heller and R. B. Clark, eds.), pp. 267–274. Academic Press, New York.

Herlant-Meewis, H. (1964). Regeneration in Annelids. *Advan. Morphogenesis* **4,** 155–215.

Hubl, H. (1956). Cited by Gabe (1966).

Kamemoto, F. I. (1964). The influence of the brain on osmotic and ionic regulation in earthworms. *Gen. Comp. Endocrinol.* **4,** 420–426.

Laverack, M. S. (1963). "The Physiology of Earthworms." Pergamon Press, Oxford.

Meglitsch, P. A. (1967). "Invertebrate Zoology." Oxford Univ. Press, London and New York.

Scharrer, B. (1937). Uber sekretorische tätige Nervenzellen bei wirbelloson Tieren. *Naturwissenschaften* **25,** 131–138.

Scharrer, E., and Brown, S. (1961). Neurosecretion. XII. The formation of neurosecretory granules in the earthworm, *Lumbricus terrestris. Z. Zellforsch. Mikroskop. Anat.* **54,** 530–540.

Simpson L., Bern, H. A., and Nishioka, R. S. (1966). Survey of evidence for neruo-secretion in gastropod molluscs. *Am. Zoologist* **6,** 123–138.

CHAPTER **9**

ARTHROPODA–CHELICERATA

The arthropod phylum is the most successful, in some respects, of all phyla because it makes up between one-half and three-fourths of the living species and its members have become adapted to extremely diversified habitats. Three main phylogenetic divisions have evolved and are often referred to as subphyla: the Trilobita which are now extinct, the Chelicerata, and the Mandibulata. The chelicerates, clearly the smaller of the two living subphyla, will be considered in this chapter. The mandibulates, divided into the aquatic (Crustacea) and terrestrial (Insecta and Myriapoda) forms will be covered in Chapters 10 and 11, respectively.

The best known examples of the Chelicerata are scorpions, spiders, and ticks. Many of these arthropods are scorned because of the harm they may cause with venoms or stings; however, the beauty and perfection of a spider's web is almost matchless in nature.

Several characteristics set this subphylum apart from the mandibulates. The body is divided into two parts: the anterior prosoma, or cephalothorax, with the head and somites bearing walking legs, and the opisthosoma, or abdomen, with reduced or modified appendages. The head is without antennae and the corresponding portion of the brain, the deutocerebrum, is reduced. No preoral appendages are present but the first postoral somite contains pincerlike chelicerae.

The nervous system of the primitive arthropod, much like that of the polychaetes, consists of a brain, circumesophageal nerve ring, and segmental ganglia comprising a ventral nerve cord. The brain in chelicerate arthropods is composed of a protocerebrum giving rise to the optic nerves, and a second division which corresponds in function to the tritocerebrum of mandibulates, providing nerves to the chelicerae. Thus, centralization has occurred in the arthropod with concentration of segmental ganglia in the forward end of the body. Variations on this theme are found throughout the phylum and especially among the Chelicerata, with the horseshoe crab retaining a considerable portion of the ventral nerve cord while the arachnids possess a nerve ring which is as compact as that of the most advanced insect.

Neurosecretory cells are located as bilateral groups in all ganglia; usually there are two groups in both the protocerebrum and tritocerebrum, and one in each of the remaining ganglia. Except in the Merostomata, axonal transport is apparent and the cells can be traced to neurohemal organs with nerve endings in association with the circulatory system. Several nonneural glandular structures, which may correspond functionally to the ecdysial gland of insects and y-organ of Crustacea, are present in the head region.

Of the three Chelicerata classes, only the Merostomata and Arachnida have been investigated by endocrinologists. The latter class is the most abundant and has received by far the most attention while the former, although offering interesting material as a living link with the Trilobites, has been studied very little.

I. MEROSTOMATA

Only three species of horseshoe crabs have been reported to contain evidence of neurosecretion. The earliest report, that of B. Scharrer (1941), is the most complete, and concerns *Xiphosura polyphemus* and *Limulus moluccanus*. She found neurosecretory cells appearing throughout the central nervous system from the cerebral ganglion to the last abdominal ganglion (Fig. 9-1). An increase in cell density from anterior to posterior regions of the circumesophageal nerve ring was apparent, but no relationship existed between the number of cells or their staining affinity and daily or yearly cycles. Morphologically, the cells appeared to be quite large, with diameters reaching 80–100 μ; they appeared to have an abundant deposit of secretory product which appeared to push the nucleus to the cell periphery. Despite the abundance of neurosecretory material, Scharrer was unable to find morphological evidence for axonal transport from the perikaryon. Similarly, in *Limulus longispina* there appears to be an

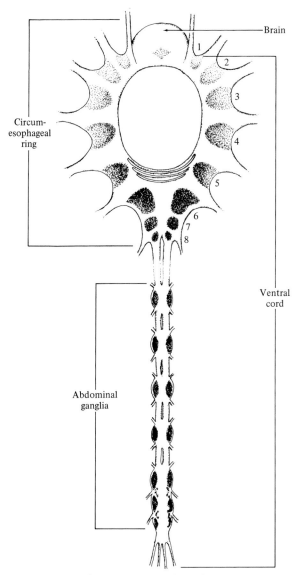

Fig. 9-1. Diagram of central nervous system of *Limulus* to show increasing numbers of neurosecretory cells anteroposteriorly around circumesophageal ring. Stippled areas represent neurosecretory cell groups; coarser data indicate greater neuroglandular activity. Ganglia of circumesophageal ring are numbered 1–8. (From B. Scharrer, 1941.)

absence of secretory product along the axon which leads from neuro-secretory cells located within the lateral rudimentary eye. However, because of considerable vascularization in the region of this eye, the presence of a neurohemal structure is a possibility, although it has not been detected in studies thus far.

Information concerning the functional importance of the described cell areas is limited to the possible presence of a chromatotropin. A substance has been extracted from the posterior portion of the nerve ring of Limulus, where the neurosecretory cells are most dense, which disperses black chromatophores and concentrates white chromato-phores in several crustaceans.

II. ARACHNIDA

A. Scorpionida, Pseudoscorpionida, Acarina

These three orders, containing scorpions, pseudoscorpions, and ticks and mites, have been investigated using recent histological techniques by Gabe (1955). In general, their morphology is similar; accordingly they will be grouped together and discussed as one unit.

Neurosecretory cells between $15-30\,\mu$ in maximum diameter are located in two symmetrically placed groups on the dorsal region of the protocerebrum. In Acarina, an additional pair is located in the anterio-ventral portion in proximity to the tritocerebrum. From all groups axonal transport of neurosecretion is easily followed through and out of the ganglion to neurohemal areas where accumulation occurs. A section of the neurohemal structure for each of the three orders is shown in Fig. 9-2A, B, C: "stomatogastric ganglion"—Scorpionida, parapharyngeal organ—Pseudoscorpionida, and "paraganglionic plates"—Acarina. The subesophageal nerve ring also contains neurosecretory cells arranged metamerically, having histological characteristics similar to the protocerebral cells. Only with con-siderable difficulty can these axons be followed, and then only to where they enter the neuropile.

No information is available concerning nonneural endocrine glands or physiological relationships, for, as Gabe states concerning all of the Arachnids and definitely these three orders, ". . . functional interpretation belongs to the future."

B. Araneae

Spiders, which are very successful small carnivores, have been the most thoroughly studied of all the Chelicerata with at least thirty genera recorded in the literature. The protocerebral neurosecretory cells are grouped in two bilateral pairs: The first is anteroventrally located and is referred to as the oral group; and the second, the aboral pair, is posterodorsal in the ganglion. One small pair is located in the anterior border of the tritocerebrum, in proximity to the oral group (Fig. 9-3). Axons from these cells can be traced through the ganglion and out of the posterior border where they converge on the retrocerebral complex composed of two paired organs of Schneider. The second or secondary organ of Schneider, discovered in 1954 by Legendre, is innervated by the pharyngeal nerve having its origin in the oral protocerebral cell group. The first or primary organ of Schneider receives its innervation by the accessary and principal nerves which originate in the perikaryon of several protocerebral cell groups (Kühne, 1959).

Both the primary and secondary organs of Schneider appear to serve as neurohemal organs since an accumulation of the neurosecretory product transported via the axon is clearly evident. Intrinsic secretory cells may also be present in both organs.

The subesophageal nerve mass contains neurosecretory cells which are associated with the ganglia of the pedipalpi, the four pairs of walking legs, and the fused abdominal nerve mass. These cells stain in similar fashion to those of the protocerebral neurosecretory cells, but the axons cannot be followed beyond the axon hillocks.

Functional significance for any of the neurosecretion in spiders is quite sketchy; however, by following the course of postembryonic development, the cells do reach their peak density and become fully charged with paraldehyde-fuschin-positive material at the time of sexual maturation. The cells then remain in this condition during copulation and oviposition, and gradually diminish thereafter. Also, during hibernation, neurosecretory material disappears and only reappears when feeding is resumed.

C. Opiliones

The harvestmen or "daddy long legs" possess two pairs of bilaterally arranged groups of neurosecretory cells: the oral and aboral

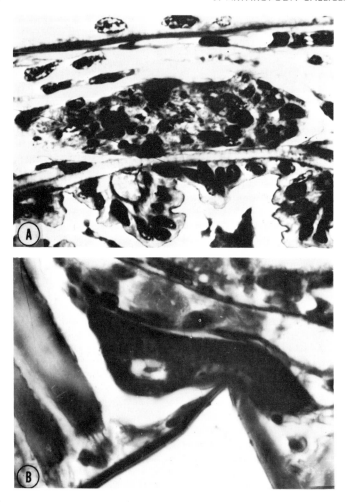

Fig. 9-2. Sections of neurohemal structure in four representations of Arachnida. A. "Stomatogastric ganglion" of *Euscorpius carpathicus,* located in the course of Police's intestinal nerve. Note storage of neurosecretory product. × 780. B. Transverse section through the parapharyngeal organs of *Garypus beauvoisi* at the level of the pharynx showing the great quantity of neurosecretory material. × 780.

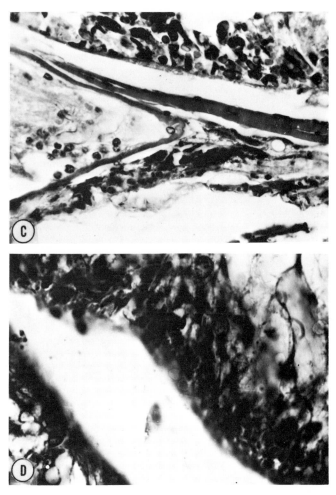

Fig. 9-2. C. Sagittal section through paraganglionic plate in *Ornithodorus lahorensis.* × 390. D. Transverse section through the posterior extremity of the cerebral ganglion of *Phalangium opilio,* showing storage of the neurosecretory material. × 780. (From Gabe, 1966.)

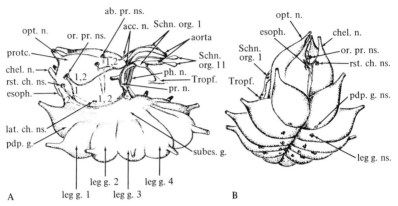

Fig. 9-3. Generalized diagram of the neurosecretory system in the brain of the spider. A. Dorsolateral view. B. Anteroventral view. 1, Neurosecretory pathways according to Gabe (1955); 2, pathways according to Legendre (1959a); acc. n., accessory nerve; chel. n., cheliceral nerve; esoph., esophagus; leg g. 1–4, leg ganglia; opt. n., optic nerve; protc., protocerebrum; pdp. g., pedipalp ganglion; ph. n., pharyngeal nerve of Schneider's organ II; pr. n., principal nerve of Schneider's organ I; subes. g., subesophageal ganglion; Schn. org. I, II, Schneider's organs; Tropf., "Tropfenkomplex." Neurosecretory cell groups: ab. pr. ns., aboral protocerebral; lat. ch. ns., lateral cheliceral; leg. g. ns., leg ganglion; or. pr. ns., oral protocerebral; pdp. g. ns., pedipalp ganglion; rst, ch. ns., rostral (oral) cheliceral. (From Bern and Hagadorn, 1965, after Kühne, 1959; Gabe, 1955; Legendre, 1959a.)

masses. Their arrangement is more dorsal in the protocerebrum than the spider's, and the maximum diameter is approximately 20 μ with centrally located nuclei. The axons from each of the four groups leave the perikaryon and course posteriorly along the esophagus, emerging from the cerebral ganglion on the posterolateral surface of the brain at the union of the protocerebral and tritocerebral ganglia. There, the two axon tracks emerging on both sides of the brain join to form a pair of paraganglionic plaques which serve as storage sites for proto-cerebral neurosecretion. The cephalic blood sinus which irrigates the posterior portion of the brain is thus in proper position to receive and transport the product throughout the body (Fig. 9-2D). A third type of neurosecretory cell appears in the posterior lateral portion of the cerebral ganglion, and only appears prominent from the time of imaginal ecdysis through the period of sexual activity. As in all other arachnids, the subesophageal nerve mass contains neurosecretory cells staining similarly to the protocerebral cells, and contains axons which cannot be traced beyond the cell body.

In the chelicerates, all information relating to function has been obtained by parallel histological studies on neurosecretory systems and observations on molting and/or reproduction (Naisse, 1959). With this same approach the conclusion may be drawn that, in the opiliones, the oral group exercises some control over the molting cycle, as there is an apparent cycle in the storage and release of neurosecretory material within these cells which coincides with each period of ecdysis. The aboral cells, on the other hand, attain their maximum size and secretory content at the time of sexual maturity.

References

Bern, H. A., and Hagadorn, I. R. (1965). Neurosecretion. In "Structure and Function in the Nervous System of Invertebrates" (T. H. Bullock and G. A. Horridge, eds.), Vol. 1, pp. 353–429. Freeman, San Francisco, California.

Gabe, M. (1955). Données histologiques sur la neurosécrétion chez les Arachnides. Arch. Anat. Microscop. Morphol. Exptl. 44, 351–383.

Gabe, M. (1966). "Neurosecretion." Pergamon Press, Oxford.

Kühne, H. (1959). Die neurosekretorischen Zellen und der retrocerebrale neuroendokrine Kimplex von Spinnen (Araneae, Labidognatha) unter Berücksichtigung einiger histologisch erkennbaren Veränderungen während des postembryonalen Lebensablaufes. Zool. Jahrb., Abt. Anat. Ontog. Tiere 77, 527–600.

Legendre, R. (1954). Données anatomiques sur le complexe neuro-endocrine retrocérébral des Araneides. Ann. Sci. Nat. Zool. Biol. Animale [11] 16, 419–426.

Legendre, R. (1959). Contribution à l'etude du systéme nerveux des aranéides. Ann. Sci. nat. (Zool.) 1, 339–473.

Naisse, J. (1959). Neurosécretion et glandes endocrines chez les Opilions. Arch. Biol. (Liege) 70, 217–264.

Scharrer, B. (1941). Neurosecretion. IV. Localization of neurosecretory cells in the central nervous system of Limulus. Biol. Bull. 81, 96–104.

CHAPTER 10

ARTHROPODA–CRUSTACEA

The principal morphological characteristics separating the chelicerates from the mandibulates, and also separating the two divisions of mandibulates, are found in the head appendages. Mandibulates possess a pair of preoral antennae plus a pair of powerful mandibles, as compared to no antennae and pincerlike chelicerae for the chelicerates. Thus the importance that taxonomists have placed on the oral appendages in naming the two subphyla is clearly evident. The first postantennal somite in the aquatic mandibulate or crustacean bears the second pair of antennae, while in the terrestrial mandibulates, insects and myriapods, this somite is reduced and has no appendages. Thus in the Crustacea the brain has three divisions: protocerebrum, deutocerebrum, and tritocerebrum, innervating the compound eyes, first antennae, and second antennae, respectively. The deutocerebrum is reduced in the chelicerates since there is no structure which corresponds to the first antenna of the crustaceans; and, since insects and myriapods do not have a second pair of antennae, their tritocerebrum is reduced. In addition the Crustacea are characterized by three body divisions (head, thorax, and abdomen) as opposed to only two in the chelicerates. Crustaceans are principally aquatic, either freshwater or marine, having sexes which are usually separate, and having compound eyes, in some subclasses, especially the Malacostraca, projecting from the body on a stalk. The latter structure becomes very important in a discussion of crustacean endocrinology since several groups of neurosecretory cells and a neurohemal structure are located within the body of the eyestalk. A popular physiological experiment

115

is to remove the eyestalks from a juvenile or adult crab and observe the adjustments in one of several physiological systems.

There are eight crustacean subclasses; only five of them have been recorded in the neuroendocrine literature. Four of the five have received only minor attention, while the Malacostraca, with two-thirds of the crustacean species, have been studied in considerable detail.

The intent of this book, set forth in the preface as that of providing an introduction to the topic of invertebrate endocrinology, must be underlined now as we begin to discuss the crustaceans as well as insects in the next chapter. A complete coverage of these two classes would extend the volume beyond the author's intent; therefore the subclass Malacostraca will be the focus of our attention and will be left intact for our discussion, with certain generalizations appearing by necessity. A monograph on the endocrinology of this order was written by Carlisle and Knowles in 1959. Five years ago Kleinholz (1965) outlined the problems existing in crustacean endocrinology in a paper which is still very significant. The reader interested in a more detailed survey of the endocrinology of Crustacea would be well advised to consult these sources, as well as the volume on neurosecretion by Gabe (1966).

I. BRANCHIOPODA, OSTRACODA, COPEPODA, CIRRIPEDIA

These subclasses have in the past been grouped with three others under the term Entomastraca, which is no longer considered taxonomically valid. They are the small Crustacea found in marine or freshwater environments, and are free-living except for some copepods and all Cirripedia.

Studies on the phenomenon of neurosecretion within the non-Malacostraca have been very limited since the first reported observations in 1957. Of the Branchiopoda, the most primitive of the modern Crustacea, neurosecretory cells have been found in the fairy shrimp and water fleas. The first group appears to lack secretory cells in the eyestalk but contains them in the anterior and posteroventral areas of the cerebral ganglion, as well as in the tritocerebrum and subesophageal ganglion. Several species lack evidence of a storage organ while a disk-shaped structure between the medulla externa and medulla interna, similar in appearance to the sinus gland, has been identified in a south Indian branchiopod. The water flea possesses four groups of neurosecretory cells in the cephalic nervous system but not in the ganglion of the ventral nerve cord. Axonal transport is difficult to detect, and no evidence is present for a storage organ.

In the Ostracoda, several types of neurosecretory cells have been detected in the protocerebral circumesophageal connectives and the subesophageal ganglion. Only limited axonal transport has been observed, with no neurohemal organs identified.

Two anterolateral groups of neurosecretory cells are found in the protocerebrum of the copepods which show axonal connections with the frontal organ. Neurosecretory products were abundant during the active summer months but showed depletion during winter diapause; this suggests a functional relationship between these cells and the annual cycle of activity.

Among the Cirripedia studied, two types of secretory cells were detected. One type was present in all ganglia of the nervous system while the second appeared in the anterior and posterior regions of ganglia of the ventral nerve cord. Axonal transport was detected in some cases, but no storage organ has been identified (Gabe, 1966).

II. MALACOSTRACA

Invariably every phylum has one or two subdivisions, either as a class, order, or family, which stand out in the minds of the most casual observers as the group illustrating all the typical characteristics of the entire phylum. The aquatic mandibulates have that distinction placed on the order Decapoda of this subclass. Shrimp, lobsters, crabs, and crayfish are found in this order. And it is a combination of the utilization of crayfish in general biology classes and the excellent cuisine quality of the first three that has made this order the best known of all Crustacea. This is also the largest and most successful of crustacean orders with approximately 8500 species. It is thus understandable that so much of the endocrine information on Crustacea has been obtained from the study of representative decapods.

The Malacostraca is a diversified subclass of mostly marine forms with the head and thorax covered by a common carapace from under which project the stalked compound eyes. The decapods extend the carapace laterally to form a branchial cavity, and also five pairs of thoracic appendages appear as leglike pereiopods.

The basic arthropod nervous system is present, with a dorsal brain, (composed of three parts: protocerebrum, deutocerebrum, and tritocerebrum), a circumenteric nerve ring, and a double ventral nerve cord with distinct ganglia.

A. Morphology

There are three component parts of the crustacean neuroendocrine system: neurosecretory cells, neurohemal organs, and nonneural endocrine glands (Fig. 10-1). Neurosecretory cells are located in the brain, portions of the eyestalk optic lobe, and within the ventral nerve cord.*

There are three neurohemal organs: (1) the sinus gland which receives axons from neurosecretory cells in the brain and on the optic lobe; (2) the postcommissure organ located immediately posterior to the esophagus and receiving axons also from the brain; and (3) the pericardial organ located in the wall of the pericardium, but variable in exact position on the wall.

Two nonneural endocrine glands have been described in Crustacea. The first is the y-organ which produces the molting hormone and is located within antennary or maxillary segments. The second is the androgenic gland located on or near the testes which influences gonadal development of the male and formation of secondary sex characteristics.

1. NEUROSECRETORY CELLS

By standard techniques, cell bodies have been identified as neurosecretory in the brain, eyestalk, and ventral nerve cord. Granular secretions in varying amounts are evident in the parakaryon of these cells and, in some cases, within axonal fibers which extend to one of the three neurohemal structures. Morphology of the cerebral ganglion varies considerably in the order Decapoda with recognition of four

---

*Gabe (1966) has suggested a change in nomenclature for the x-organ and associated areas in the eyestalk. He proposes that the sensory pore x-organ be called the "organ of Bellonci" in honor of the Italian zoologist who discovered the structure in the nineteenth century. Also the ganglionic x-organs should be named the "organs of Hanstrom" in honor of the Swedish histologist who discovered the protocephalic neurosecretory pathway in arthropods. Justifications for these changes are printed in detail by Gabe (pp. 203–215). Although points of confusion certainly do exist, the choice to stay with the most familiar nomenclature is made in this case since the advantages of going with the new as proposed by Gabe are not judged sufficient enough to offset the disadvantages naturally incurred in such a change. Name changing is a slow process; it should have the support of the researchers within the discipline concerned, and not be handed down by textbook writers.

general types (Fig. 10-2). In light of this variation of gross structure, the pattern of neurosecretory cell location is not constant, and homology is made difficult among the different cell groups.

Four cell types located in the eyestalk and cerebral ganglion have been designated in the crayfish *Orconectes* (Fig. 10-3). The largest

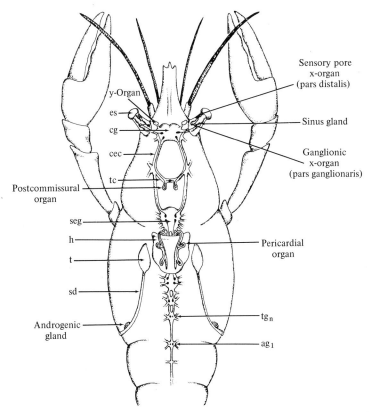

Fig. 10-1. The endocrine system of a generalized male crustacean. Principal endocrine areas are indicated on the figure with complete labels; however, neurosecretory cells are found throughout the central nervous system, and the x-organ in the eyestalk represents only one major concentration of them. The endocrine system includes neurohemal organs such as the sinus gland, postcommissural organ, pericardial organ, and nonneural epithelial glands such as the y-organ and androgenic gland. In the central nervous system, the nuclei of neurosecretory cell bodies are indicated as closed circles, and their axonal terminations as open circles. Abbreviations: es, eyestalk; cg, cerebral ganglia; cec, circumesophageal connective; tc, tritocerebral commissure; seg, subesophageal ganglion; h, heart; t, testis; sd, sperm duct; ag_1, first abdominal ganglion; tg_n, last thoracic ganglion. (From Gorbman and Bern, 1962.)

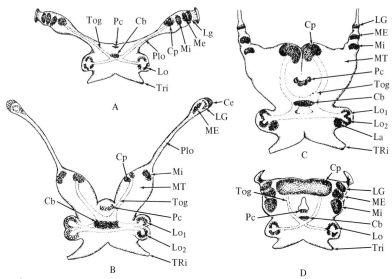

Fig. 10-2. Diagram of morphological types of cerebral ganglia in decapod Crustacea. A. *Carcinus*-type; B. *Emerita*-type; C. *Calocaris*-type; D. *Athanas*-type. cb, Central body; ce, compound eyes; cp, corpora pedunculata; la, lobus accessorius; lg, lamina ganglionaris; lo, lobus olfactorius; me, medulla externa; mi, medulla interna; mt, medulla terminalis; pc, pons cerebri; plo, pedunculus lobi optici; pc, ponc cerebri; tog, tractus olfactorio-globularis; tri, tritocerebrum. (After Gabe, 1966, from Hanström, 1947.)

cell has a maximum diameter of 40–60 μ and shows a tendency for aggregation of neurosecretory material. The second type ranges from 20–30 μ but often carries the aggregation further, with apparent fusion of the neurosecretory material into clumps of 1–4 μ in diameter. Type three differs from the preceding class in only size and position in cytoplasm of the vacuoles, while type four is much the smaller cell and shows no aggregation or clumping of neurosecretory material.

In several species of crabs, three cell types have been distinguished. The largest and smallest types have some similarities to types 1 and 4 of *Orconectes*, but the crab cell types are distinguished less in size than in the differential staining affinity with Mallory's triple stain and chrome-hematoxylin of the Gomori method. To illustrate an extreme case of classification in the Brachyura, Matsumoto (1959) devised a scheme for classifying neurosecretory cells into eleven categories on the basis of size, location, and staining (Table 10-1). A schematic map showing the generalized location of neurosecretory cell areas in crabs is provided in Fig. 10-4.

Neurosecretory cells have also been found in the subesophageal, thoracic, and abdominal ganglia of most decapods. Their appearance differs little from the cells found elsewhere in the central nervous system, and as many as five types of cells have been described.

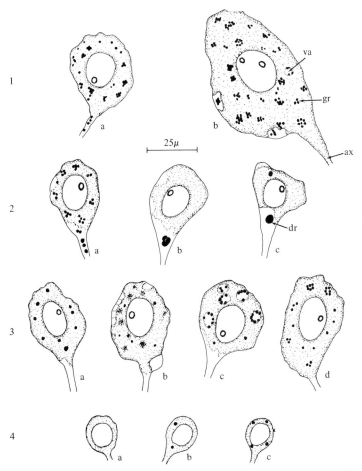

Fig. 10-3. Diagrammatic representation of neurosecretory cell types in eyestalk and cerebral ganglion of *Orconectes virilis*. 1–4, Cell types; a–d, stages of the secretory cycle; ax, axon; dr, droplet; gr, granule; va, vacuole. (From Gabe, 1966, from Durand, 1956.)

TABLE 10-1
Morphological Features of Neurosecretory Cells in *Brachyura*[a]

Type	Size	Site	Axonal transport	Local discharge
A	Large	Thoracic ganglia	CHP+[b]	CHP+ or −
A	Large	Commissural ganglia	FP+ or −[c]	FP+ or −
A	Large	Cerebral ganglia	FP+ or −[c]	Vacuoles
A'	Large	Thoracic ganglia	CHP+, FP−	—
E	Large	Eyestalk	CHP+, FP−	—
B	Small	All ganglia	CHP+, FP+ or −	—
B'	Small	Thoracic ganglia	—	CHP+, FP−
C	Small	Thoracic ganglia	CHP+, FP−	—
α	Small	Eyestalk	CHP+, FP+ or −	CHP−
β	Small	Eyestalk	CHP+, FP+	—
δ	Small	Eyestalk	—	CHP+, FP−
E	Very small	Cerebral ganglia	CHP+, FP−	—
γ	Very small	Eyestalk	CHP+, FP+	—

[a] From Gabe (1966), after Matsumoto (1959).
[b] CHP, chrome-hematoxylin.
[c] FP, paraldehyde-fuchsin.

The general arrangement of neurosecretory cells in the eyestalk has attracted the attention of many investigators. The x-organ (Gabe: organ of Hanström) is the main neurosecretory body on the optic lobe, and is most consistently located on the medulla terminalis. Variation occurs as similar groups of cells may be found on the medulla interna and/or medulla externa (Fig. 10-5). Axons have been traced from the various cell groups and, with only one exception, terminate in the sinus gland. The exception is the possibility that axons from the medulla terminalis may extend distally entering the sensory pore x-organ, but considerable debate remains on this point.

2. NEUROHEMAL ORGANS

The sinus gland in the eyestalk is probably the most constant component of the malacostracan neuroendocrine system, and is easly recognizable in a living specimen because of its blue-white opalescence. The gland is a thick dish-shaped structure on the neurolemma of the medulla terminalis; it is separated from the blood by a thin membrane. There may be invagination of this membrane into the gland to provide greater surface area for hormone diffusion. Electron microscopy has revealed that many of the axons are arranged perpendicularly to the blood–gland membrane. Axons entering this gland originate principally in the x-organ but may also arise in other regions of the central nervous system. Therefore the bulk of the gland consists of the aggregation of axon terminals, with only a small complement of connective tissue or glial cells. The suggestion that intrinsic neurosecretory cells are a part of the gland has not been supported, but the influence of storage at axonal terminals on the chemical nature of the neurosecretion is a possibility. Indeed, histochemical tests have indicated that such modifications are quite probable. The suggestion has also been made by Potter (1958) that the six cell types found in the x-organ can be related to the six axonal types observed

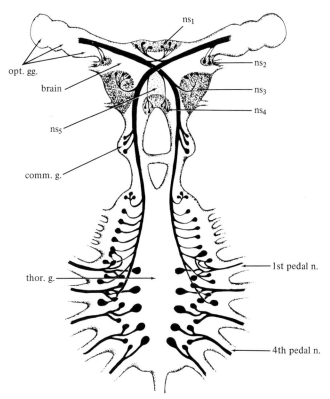

Fig. 10-4. Generalized diagram of neurosecretory pathways in crabs, omitting the eyestalk system. Fibers from the entire central system extend into the eyestalk while only fibers from the thoracic ganglion enter the pedal nerves. ns_1–ns_5, Cerebral neurosecretory cell groups; ns_5, indicated by hatching, is located ventrally; comm. g., commissural ganglion; opt. gg., optic ganglia; thor. g., thoracic ganglion; pedal n., pedal nerves. (From Bern and Hagadorn, 1965, after Matsumoto, 1958.)

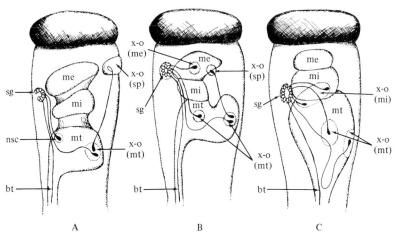

A   B   C

Fig. 10-5. Diagrams of examples of crustacean eyestalk neurosecretory systems. Dark ovoids represent neurosecretory nuclei; light ovoids represent neurosecretory terminals. A. Shrimp, *Lysmata seticaudata*. Secretory structures, largely of unknown significance, are associated with the sensory pore x-organ. Neurosecretory neurons proceed from the brain and x-organ to the sinus gland. B. Shrimp, *Leander serratus*. X-organs are present in the medullae terminalis and externa, and their fibers terminate in the sinus gland. Although a sensory pore is absent in this species, the sensory pore x-organ is present. C. Crab, *Gecarcinus lateralis*. No sensory pore x-organ is present. bt, Tract of neurosecretion-bearing axons from brain and other parts of central nervous system; me, mi, mt, medullae externa, interna, and terminalis of eyestalk ganglia, respectively; nsc, isolated neurosecretory cells; sg, sinus gland; sp, sensory pore; x-o, x-organ. (From Gorbman and Bern, 1962, after Carlisle and Knowles, 1959.)

in the sinus gland. Although electron microscopy has been limited in its application to crustacean endocrinology, Bunt and Ashby (1967) disclosed that five different types of neurosecretory granules, homogenous within each axon in respect to dimension, shape, and electron density are present in the sinus gland of the crayfish, *Procambarus clarkii* (Fig. 10-6, Table 10-2). Attempts to experimentally alter the granule population by changing environmental parameters may aid in correlating granule types with assayable physiological functions.

The third eyestalk structure of interest is the sensory pore x-organ (Gabe: organ of Bellonci). This consists of bipolar nerve cells in contact with the sensory pore exoskeleton and connected to the medulla terminalis by way of a nerve fiber. There is currently considerable debate in interpreting the histology of the organ. The controversy is over whether it is a sensory organ with afferent nerves extending to the x-organ, or a neurohemal structure receiving neurosecretory axons from the x-organ, or possibly a combination of both.

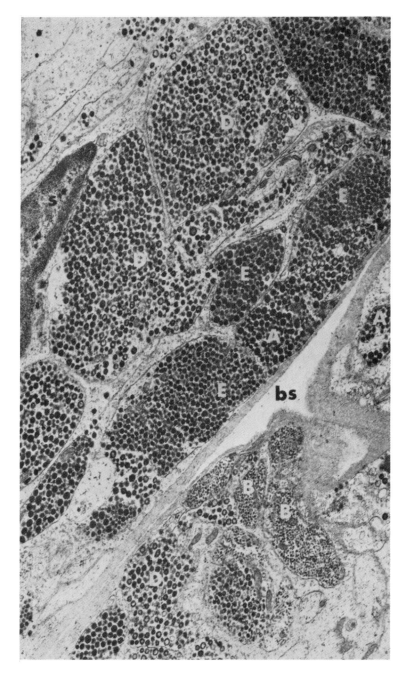

Fig. 10-6. A section of the sinus gland from *Procambarus clarkii* after prolonged staining with lead tartrate. Types A, B, and E granules are densely stained, while type D granule cores are washed out of the section. bs, Blood sinus; s, supporting cell nucleus. × 21,600. (From Bunt and Ashby, 1967.)

TABLE 10-2
Characteristics of the Five Neurosecretory Granule Types Found in the Sinus Gland of
Procambarus clarkii[a]

Granule type	Diameter (Å)	Shape	Electron density	Staining characteristics
A	2000–2400	Round, oval, or tear-drop	Very dense	Stains densely with uranyl acetate (U.A.) or lead tartrate (L.T.)
B	500–900	Round or oval	Moderately dense	Densely with U.A. or L.T.
C	1000–1200	Round	Very dense	Densely with U.A. or L.T.
D	1500–1700	Round, oval, or irregular	Moderately dense, often with clear halo between core and limiting membrane	Densely with U.A. or brief exposure to L.T. but granule core washes out after prolonged staining with L.T.
E	2000–2400	Round or oval	Moderately dense	Moderately dense with U.A. or L.T.

[a] From Bunt and Ashby (1967).

The postcommissural organs are found posterior to the esophagus and attached to the circumesophageal connective in all decapods except the Brachyura. They appear on fresh specimens as two small bluish-white swellings on the postcommissural nerves which extend from the tritocerebral commissures to a pair of muscles associated with the exoskeleton (Fig. 10-7). Axons from neurosecretory cells believed to be in the tritocerebrum pass down the circumesophageal connectives into the tritocerebral commissure and there bifurcate, sending one branch into each of the two postcommissure nerves. The axons, on reaching the organ, split into branches which ramify throughout the structure with the bulbous endings contacting the epineurium, thereby allowing easy diffusion of the neurosecretory material from the axons into the hemolymph.

Effector nerves also pass from the central nervous system along the same route but continue through the postcommissural organs to innervate the muscles attached to the cephalic apodemes of the endophragm and to the dorsal hypodermis.

The last of the neurohemal structures are the pericardial organs, originally known as the dorsal lamelae, which are located within, or attached to, the lining of the pericardial cavity. In *Squilla mantis* they are easily observed in the living specimen, about 1 mm in length and 0.5 mm in depth. The organs receive numerous axons from neurosecretory cells in the thoracic as well as other segmental ganglia, and

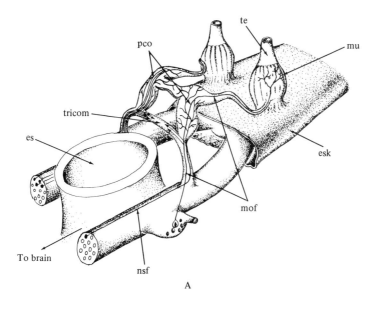

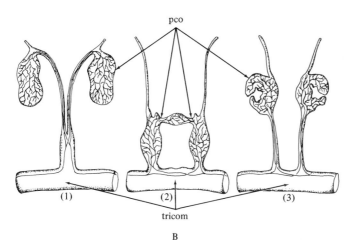

Fig. 10-7. A. Semidiagrammatic view of the postcommissure organs of *Leander serratus*. Each circumesophageal connective contains four neurosecretory fibers which extend to the postcommissure organ, and two motor fibers which traverse each organ to innervate the muscle. Only one fiber of each type has been shown in the left connective commissure. B. Variation in the position of the postcommissure organs of three Crustacea: (1) *Penaeus;* (2) *Leander,* and (3) *Squilla.* c, Circumesophageal connective; esk, endophragmal skeleton; mof, motor fiber; mu, muscle; nsf, neurosecretory fiber; pco, postcommissure organs; es, esophagus; te, tendon; tricom, tritocerebral commissure. (From Carlisle and Knowles, 1959.)

probably the subesophageal ganglion. Unlike the sinus gland and the postcommissural organ, this structure possesses intrinsic neurosecretory cells which resemble similar cells in the insect's neurohemal structure, the corpus cardiacum. Two other cell types are observed. Schwann cells ensheath axons leading into and through the center of the organs but do not continue to the terminal portion of the axons, and supporting or glial-like cells are located only in the peripheral region of the structure.

According to the work of Knowles (1962) on the pericardial organ of *Squilla mantis,* the axon terminals from types A and B neurosecretory fibers lie directly below the surface of the organ and are separated by a layer of an amorphous, collagen-containing membrane from the pericardial sinus. Axons of type A contain ovoid, electron-dense neurosecretory granules, with a diameter approximately 1500 Å. Granules of B axons are less regular in shape, having an average diameter of 1200 Å, and are clearly surrounded by a distinct membrane which is often separated from the contained opaque material (Fig. 10-8).

3. NONNEURAL ENDOCRINE ORGANS

Some 15 years ago an epithelial gland, the y-organ, was described in a large number of Malacostraca and was shown to have an influence over the crustacean molting process. The gland appears to be quite similar to the insect's ecdysial gland which produces ecdysone, the insect molting hormone. This is to be expected, for both are derived from ectodermal ventral glands. The location of the pair of y-organs seems to be dependent upon the position of the excretory organs; if the location of excretory organs is on the maxillary segment, then the y-organs will be found on the antennary segment, while the opposite arrangement is also quite possible. In either position the endocrine glands are innervated by the subesophageal ganglion. The appearance of the y-organs varies from group to group; in the Brachyura they are conical, in the Natantia, lenticular, and in Isopoda and Amphipoda, foliaceous; however, the histology is more uniform, with small cells of equal size averaging 10 μ in diameter.

Fig. 10-8. A transverse section through a small portion of the cortical region of the pericardial organ of *Squilla mantis.* The fibers are closely packed, not ensheathed by Schwann cell cytoplasm, though separated in places by supporting cells (c). Two fiber types are shown (a and b). × 11,400. (From Knowles, 1962.)

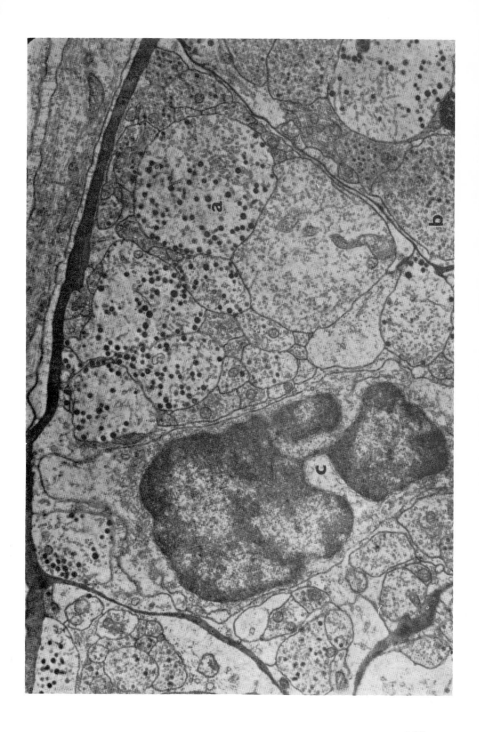

129

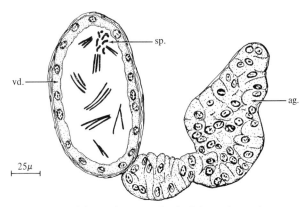

Fig. 10-9. Section of the androgenic gland of the male *Orchestia gammarella*. vd, Vas deferens; sp, sperm; ag, androgenic gland. (From Charniaux–Cotton and Kleinholz, 1964.)

The second nonneural endocrine organ is the androgenic gland attached to the distal region of the vas deferens (Fig. 10-9). From this gland flows a hormone, believed to be proteinaceous, which is responsible for the differentiation of all primary and secondary male sex characteristics. In *Orchestia gammarella*, where the gland was first discovered in 1954 by Charniaux-Cotton, it appears as a pyramidal mass of cells about 250 μ across the base, on the surface of the genital tract. Often smaller masses of glandular tissue are found proximally to the main structure. Microscopically, the gland may vary considerably in form, appearing as a vermiform structure in *Carcinus* about 7 mm long, 35 μ wide, and entwined about the vas deferens.

B. Physiology

Physiological processes of the Crustacea which have been found to respond to endocrine regulation have been classified into three groups: kinetic, morphological, and metabolic (Carlisle and Knowles, 1959). Each then can be broken into several subtopics; for example, kinetic processes into somatic pigmentation, retinal pigment migration, and cardiac regulation. An interesting diagram illustrating nine processes under hormonal influence was drawn by Charniaux-Cotton and Kleinholz (1964), and is reproduced in Fig. 10-10. An arrangement which is between these two extremes will be utilized in this chapter.

Although the difficult problems of neurosecretory cell morphology and classification confronting the molluskan and annelidan endo-

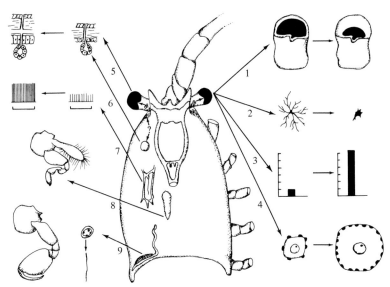

Fig. 10-10. Summary of hormonal functions in crustaceans. On the right side, 1–4 represent hormones from the eyestalk; on the left, 5–9 represent eyestalk and other endocrine effects. 1, Light-adapting distal retinal pigment hormone; 2, chromatophoro-tropins; 3, hyperglycemic hormone; 4, eyestalk ablation results in ovarian growth through precocious vitellogenesis; 5, molt-inhibiting hormone of the eyestalk, probably acting normally on the y-gland; 6, the y-gland, from which is secreted a molting hormone; 7, the pericardial organ, extracts of which accelerate the heart rate; 8, ovarian hormone regulating female secondary sexual characters; 9, androgenic gland of male, regulating spermatogenesis and secondary sex characters in male. (From Charniaux–Cotton and Kleinholz, 1964.)

crinologists are not present with the crustaceans, the possibility of overclassification of eyestalk extracts does exist and is therefore a temptation which must be guarded against. Barrington (1964) commented very appropriately on this matter by referring specifically to the chromactivating hormones: "There is, however, a real danger of begging important questions of crustacean endocrinology by too readily regarding an ever-increasing diversity of imperfectly character-ized substances as representing so many distinct chromactivating hormones."

1. SOMATIC PIGMENTATION

The ability of an organism to alter its coloration in response to some intrinsic or extrinsic stimulus is found in representatives of both major divisions of the animal kingdom. In the invertebrates the decapod Crustacea are well known for their diversity of color changes. Some of these changes are slow and predictable since they are in response to changes in the position of the sun, moon, and earth; others are more rapid in response to a change in the background coloration, or to an alteration in illumination. Under increased illumination with higher temperature, dark pigments are often concentrated and white pigments dispersed, thus allowing greater light reflection and a retention of a more optimum body temperature.

Changes in crustacean somatic pigmentation illustrate a phenomenon which is referred to as physiological color change, dependent upon the movement or redistribution of a given amount of pigment rather than a morphological color change, which refers to an adjustment in the total amount of pigment in a body. The basic unit producing these changes is the chromatophore which may refer to a single pigment cell, but more often to a multinucleated syncytial body located in the hypodermis. Each cell contains only one pigment, i.e., red, yellow, brown, or white, but in a given syncytial body a number of pigments may be available. Such color changes would require sufficient regulatory mechanisms to allow each pigment to adjust independently or in combination with others. Shrimp illustrate just such an elaborate system with chromatosomes and polychromatic chromataphores containing as many as four different pigments within one syncytial complex. Thus the crustacean does not alter the size of the chromatophore but, rather, the location of the pigment granule within the chromatophore. When the pigment is concentrated near the center of the cell, a minimum display of color is given; when dispersed throughout the cell, a maximum display is obtained (Fig. 10-11).

In contrast to the cephalopods which regulate their chromatophores by excitatory and inhibitory nerves, the crustacean depends upon blood-borne regulatory agents, the chromatophorotropins, for proper color adjustment. This distinction has not always been apparent. In 1876 Pouchet discovered that when the eyestalks from shrimp were removed, their ability to adjust body coloration to background conditions was lost. The conclusion drawn from this experiment was that color change in shrimp was under nervous control which was activated by the photoreceptors on the eyestalk. In 1928 Koller and Perkins separately showed that chromatophores of blinded shrimp were responsive to an injection of blood and eyestalk extract from normal shrimp, and that nerve connections played no part in this adjustment.

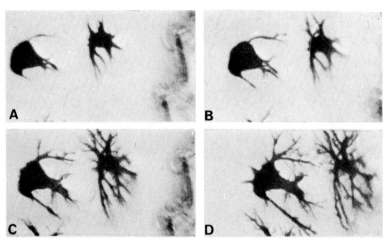

Fig. 10-11. Successive stages (a–d) in the pigment expansion or dispersion of two red chromatophores from the uropod of the shrimp *Palaemon*. The cells are of a fixed size and shape, and the pigment within them is only redistributed. (Knowles, 1955.)

A classic experiment by Koller was to adapt shrimp to black, yellow, or white backgrounds, and bring about the dispersion of black pigment when blood from the black-adapted shrimp was injected into shrimp on a white background. However, it was not until the concept of neurosecretion was established some time later that the emphasis was correctly placed on the control of somatic color by the neuroendocrine tissue. These hormonal agents therefore are neurosecretions which originate in the x-organ of the eyestalk, or in comparable cells in the cerebral, thoracic, or even abdominal ganglia. They may then be transported to, and stored within, the sinus gland or the postcommissural organ to be released upon proper stimulation, or the substances may be released to the blood at some unidentified area closer to its source. Herein lies one of the major problems in understanding the endocrinology of crustacean chromatophores, for there may be multiple hormones released to the blood from a number of different sources, some of which may still be unidentified. Once the chromatophoratropins are in the blood they may act individually, in concert, or even synergistically to one another on the appropriate chromatophores (Fig. 10-12).

Questions concerning the number of hormones utilized for color adjustment, the similarity of chromatophorotropins between species, the chemistry of the hormones, and their mechanism of action are all

Fig. 10-12. The common reddish-brown European shrimp, *Leander serratus,* showing
various patterns of somatic pigmentation following injection of salt water extracts from
two neurohemal areas. A. Normal pale animals; D. darkened animal, whose eyestalks
had been removed 4 weeks previously; B. an eyestalkless animal which had received an
injection of an extract from two sinus glands (note the paling as compared with D,
resulting largely from concentration of small red chromatophores of the body and tail,
and partly from concentration of large red chromatophores); C. an eyestalkless animal
which has received an injection of an extract from two postcommissural organs (note
that the paling response is different from that seen in B). The small red chromatophores
are little affected, but the concentration of large red chromatophores is largely responsible for the paling that has occurred. (From Gorbman and Bern, 1962, rephotographed
from Knowles, 1955.)

difficult to answer, and much of the current research deals with
these problems. One of the simpler conditions is found in the isopods
where the organism exists in either a lighter or darker phase, dependent
upon the environmental conditions which, in turn, control the condition of the monochromatic chromatophore. This phenomenon
could be easily regulated by varying concentrations of two antagonistic
neurosecretions, a pigment-concentrating and a pigment-dispersing
factor. There is evidence that this is the probable explanation (Fingerman, 1965). On the other hand, natants, which contain polychromatic chromatophores with three or four different pigments in each
complex, represent a more complicated system. Postulating the
control of such a system requires a multiple-hormone hypothesis.

According to this a number of hormones would control the response of various chromatophores and could evoke individual or combined chromatophore response which would be maximal or graded in degree. An example in a two-chromatophore system is the crayfish *Cambarellus shufeldti* which has red and white chromatophores. Here the degree of pigmentation is dependent upon the absolute concentration of the eyestalk chromatophorotropins in the blood, as well as the ratio of the pigment-dispersing and pigment-concentrating hormones. In addition there is evidence that the chromatophores may be dependent upon an additional factor produced by the postcommissural organ (Fingerman, 1965).

As to the chemical nature of two chromatophorotropins, the concentrator of pigment in palaemonid erythrophores (ECH), and the disperser of brachyuran melanophore pigment (MDH), indications are that they are polypeptides. This is supported in part by the fact that hormone inactivation occurs in the presence of pepsin, trypsin, and chymotrypsin (Table 10-3).

2. RETINAL PIGMENT MIGRATION

A less conspicuous type of pigment migration than that of changes in the somatic coloration is found within the crustacean eyes. As in all arthropods, the compound eye is composed of many individual, tubular, photoreceptive units known as ommatidia. Each is composed of a series of distal lenses and several proximal retinula cells which together form the rhabdom on which the transduction of light energy into nervous stimulation occurs. Crustacea are able to adjust the amount of light impinging upon the rhabdom by altering the location of light-absorbing or light-reflecting pigment within each ommatidium. This internal adjustment is in response to the intensity of light falling upon the eye.

There are three sets of eye pigment: distal pigment, located in the distal pigment cells; reflecting pigment, in the tapetal cells that lie between the proximal ends of adjacent ommatidia; and a proximal pigment, found in the photoreceptive retinular cells (Fig. 10-13). In bright illumination the ommatidium adjusts to a light-adapted condition with the sides of the rhabdom shielded by the black, distal, and proximal pigment, and only light rays that enter along the axes of the ommatidia stimulate the rhabdom. In dim illumination the eye becomes dark-adapted; the distal pigment moves distally to surround the dioptic apparatus, the proximal pigment moves below the base-

TABLE 10-3
Properties of Five Possible Crustacean Eyestalk Hormones[a]

	DRPH[b]	ECH[c]	MDH[d]	DH[e]	MIH[f]
Solubility in:					
100% Ethanol	i[g]	s[h]	s	—	s
<100% Ethanol	s	s	s	—	—
Ethyl ether	—	i	i	—	i
Acetone	i	s	sl.[i]s in 90%	s in 50%	i
Petroleum ether	—	—	i	—	—
Stability at or in:					
100°C	stab	stab	stab[j]	destr	—
20°C	—	—	destr[k] slowly	—	—
0.1–0.3 N HCL	—	stab	stab	—	—
0.1–0.3 N NaOH	—	stab	stab	—	—
6 N HCl	—	destr at 100°	—	destr at 4°	—
Inactivated by:					
Tissue extract	Yes	Yes	Yes	—	—
Chymotrypsin	Yes	Yes	Yes	Yes	—
Papain	—	Yes	Yes	—	—
Pepsin	—	Yes	—	Yes	—
Trypsin	Yes	?[l]	Yes	Yes	Yes
Dialyzable through cello-phane or cellulose	Yes	Yes	Yes	none or sl.	—

[a] From Kleinholz (1966).
[b] DRPH, retinal pigment hormone.
[c] ECH, concentrator of erythrophore pigment.
[d] MDH, disperser of melanin pigment.
[e] DH, diabetogenic hormone.
[f] MIH, molt-inhibiting hormone.
[g] i, insoluble.
[h] s, soluble.
[i] sl., slight.
[j] stab, stable.
[k] destr, destroyed.
[l] ?, conflicting or uncertain observation.

ment membrane, and the reflecting white pigment takes its position around the photosensitive area, thereby increasing the amount of light reaching the rhabdom.

Movements of the distal and reflecting pigments are partially under hormonal control, as can be shown by several experimental procedures. When an extract is prepared from eyestalks of light-adapted shrimp and injected into dark-adapted recipients, the distal pigment moves proximally to shield the rhabdom, while reflecting pigment leaves the vicinity of the retinal cells and moves below the basement membrane; both adjustments are characteristic of the light-adapted eye. No change, however, is noticed in the proximal pigment. Further evidence for the presence in the eyestalk of a light-adapting hormone is that the distal pigment moves permanently into a dark-adapted position on the extirpation of the sinus glands. Limited evidence is available suggesting the presence of a dark-adapting hormone in the eyestalk, but it needs further documentation. There is no indication of how the position of the proximal pigment is regulated, whether it is under hormonal or nervous control.

Available biochemical data on the distal retinal pigment hormone (DRPH) places it in a similar category to that of the chromatophorotropins, i.e., a polypeptide of low molecular weight (Table 10-3).

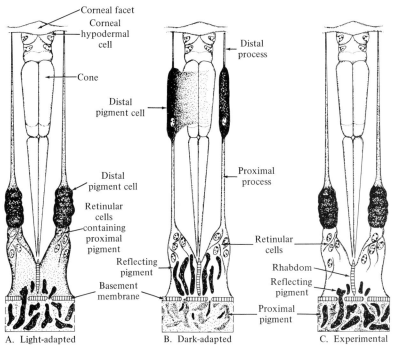

Corneal facet
Corneal hypodermal cell
Distal process
Cone
Distal pigment cell
Distal pigment cell
Proximal process
Retinular cells containing proximal pigment
Reflecting pigment
Retinular cells
Rhabdom
Basement membrane
Reflecting pigment
Proximal pigment

A. Light-adapted   B. Dark-adapted   C. Experimental

Fig. 10-13. Ommatidia from the eyes of the shrimp *Palaemonetes vulgaris*, showing general morphology and the positions of the pigments under different conditions. A. The light-adapted state; B. the dark-adapted state; C. an ommatidium from an animal which had been adapted to darkness and then had an injection of an eyestalk extract from light-adapted animals. The distal and reflecting pigments assume the light-adapted position, while the proximal pigment is not influenced. (From Turner, 1966.)

3. CARDIOACCELERATION

Upon proper nervous stimulation the pericardial organs release, into the surrounding fluid, an active substance which increases rate and amplitude of the heartbeat. This was suggested in the early investigations on this structure when its topographic relation with the circulatory system was made apparent, and has now been supported experimentally in a number of *in vitro* profusion studies.

Utilizing this experimental approach, Cooke (1964, 1967) has provided valuable information on crustacean neuroendocrinology. When an isolated pericardial organ of the spider crab *Libenia emarginata* is stimulated adequately, measurable quantities of this heart-excitatory material appear in the bathing fluid within 6 minutes. Cooke has shown that a correlation exists between the maximum release of the active neurosecretory material and the appearance of compound action potentials recorded from the pericardial organ trunks (Fig. 10-14). After analyzing the qualitative nature of this relationship more completely, Cooke (1967) concluded that ". . . the electrical activity which is correlated with release of neurosecretory material represents action potentials conducted by the axons of the neurosecretory cells, and that neurosecretory material is released

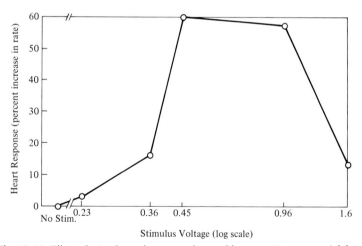

Fig. 10-14. Effect of stimulus voltage on release of heart-excitatory material from a pericardial organ. The graph plots heart responses against the log of the stimulus voltage. A large amount of excitatory material appeared in the bathing fluid when the stimulus voltage was sufficient to give a large component of slowly conducted activity in the compound action potential (From Cooke, 1967.)

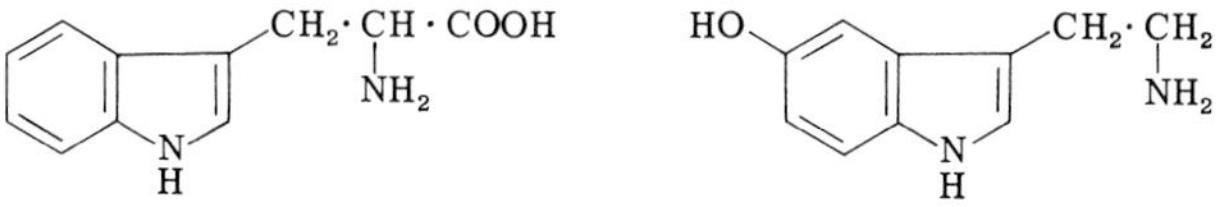

Figure. 10-15. The amino acid tryptophan, and one of the three hydroxyindoles which are possibly secreted from the pericardial organ.

from the terminals as a result of these action potentials." This is an excellent *in vitro* technique which offers the invertebrate physiologist a method for providing significant information relevant to questions of neurosecretory transport and release.

As to the chemical nature of this neurosecretory material, much has been written and the final answers are still awaited. All of the analyses thus far indicate that the active material may either be a hydroxyindole or a polypeptide. The details of the argument for and against the two candidate groups will not be presented here, but are well presented in the two articles by Cooke (1966) and Belamarich and Terwilliger (1966). The three hydroxyindoles which have been discussed are 5-hydroxytryptamine (5-HT), 6-hydroxytryptamine (6-HT), and 5,6-dihydroxytryptamine (5,6-HT), which are all variations of the amino acid tryptophan (Fig. 10-15), and are usually considered neurotransmitters rather than neurohormones. Concerning the second group, there appear to be two similar polypeptides with molecular weight between 700–1500 which retain activity after chromatographic separation. The amino acid residue, following hydrolysis, has reacted positively to tests for glutamyl, aspartyl, lysine, glycine, alanine, and either serine or valine. Whether there are normally two active polypeptide molecules, or one compound which is broken during the extraction process, has not been resolved. Nevertheless the indications now are that the neurosecretory material is a peptide rather than an hydroxyindole.

An additional physiological process, the regulation of the respiratory system, may be under the control of this pericardial organ and its anterior extensions. The extensions are so arranged that their secretions flow into the veinous blood which reaches the respiratory muscles and gills. In this way such material could influence either the rate of pumping of the respiratory chamber or the peripheral resistance of the branchial blood vessels. But insufficient data limit any further implications concerning this possible relationship.

4. MOLTING

The life of an arthropod consists of successive periods of apparent growth when the animal escapes from its limiting, hard exoskeleton, and expands its new, soft cuticle to provide room for subsequent body enlargement. This sloughing off the old cuticle may be referred to as either molting or ecdysis. From an external view, growth could then be considered a discontinuous process with an increase in size restricted to the period between the loss of the old exoskeleton and the expansion and subsequent hardening of the new. Internally, growth would be more continuous with the morphological and physiological activities, associated with all division and energy metabolism, more correctly extending from one molt to the next, except in periods of inactivity.

The crustacean molt cycle can be most easily divided into four divisions: premolt, molt, postmolt, and intermolt. During the premolt, oxygen consumption increases, glycogen is deposited in the hypodermis, and reusable nutrients are reabsorbed from the old exoskeleton. This is a time for limb regeneration and often morphological modifications of the hypodermis. At the end of the premolt, the old cuticle, which has been weakened by enzymic action, splits, thereby facilitating the actual molt. Since the animal is quite soft and defenseless as it emerges, this period is the most vulnerable in the crustacean's life cycle. By excessive water intake, in the immediate postmolt period, the body expands its hardening exoskeleton, thus creating space for subsequent body enlargement. The crustacean may then enter an intermolt period which might extend for several months, coinciding with winter inactivity or hibernation, or may just be characterized by a short period of growth stability at any time during the year. Many crustaceans that have reached puberty utilize this period for reproduction, whereas most juveniles omit the intermolt, or at best incorporate only a brief rest between successive molts.

The regulation of these periods of seasonal activity are under neuroendocrine control. Although the timing and duration of such periods vary according to geographical locale, environmental conditions, age, and sex of the arthropod, the success of each organism depends upon the efficiency of his neuroendocrine link connecting the nerve cells with the hormonally activated target organ (Fig. 10-16).

There are a number of external afferent factors which influence the crustacean growth. One of the more important is light, its intensity as well as its photoperiod. For example, when the land crab *Gecarcinus*

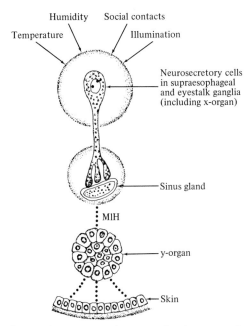

Fig. 10-16. Molting in Crustacea is under a second-order neuroendocrine mechanism with a neurosecretory molt-inhibiting hormone (MIH) supressing the activation of the y-organ. Upon the proper environmental stimuli the MIH is reduced in concentration, thereby allowing the y-organ to synthesize and release one or several molting hormones (MH) which control the epidermal cell's activity. (From Scharrer, 1958.)

lateralis is kept under constant illumination, molting is inhibited, yet in other species prolonged darkness will produce equal results. Temperature will also influence the duration of the intermolt, lengthening it when the environmental temperature either goes below or above a certain point. However, this stimulus may influence the rates of metabolic reactions more directly than it may influence the neuroendocrine system. Salinity, humidity, food availability, and lack of privacy are other factors on which the initiation and continuance of molting seem to be dependent.

The endocrine basis for the control of molting was not demonstrated until after the basic endocrine mechanism of chromatophore regulation was understood; this coincided with the discovery of the eyestalk as an endocrine organ. Since that time substantial evidence has led to the hypothesis that in adult malacostracans molting is under the control of at least two hormones: a molt-inhibiting hormone (MIH)

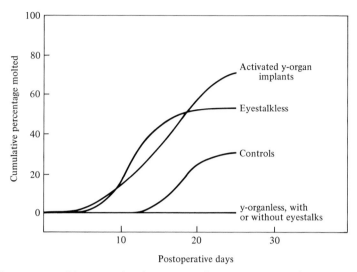

Fig. 10-17. Modifications of molting in juvenile *Carcinus maenas* by experimental means. Molting ceased when the y-organs were removed; the level of the controls was doubled with the removal of eyestalks and molting was stimulated still further when several-y-organs from eyestalkless donors were implanted. (From Passano, 1961.)

produced by the x-organ, and a molt-inducing hormone (MH) secreted by the y-organ, or molting gland. The possibility that the same mechanism, or one similar, is functional in the immature crustacean has been discussed recently by Costlow (1968). An endocrine demonstration which has been repeated often on many different crustaceans but primarily on crabs is to induce molting by the removal of both eyestalks and then inhibit molting by the injection of eyestalk extracts into the body. If, however, the y-organs are absent when the eyestalks are removed, molting does not occur, but if the y-organs are replaced in organisms without eyestalks, the ecdysial processes will be initiated (Fig. 10-17). This evidence strongly suggests that the y-organ, through the production of the MH, is the principal molt-inducing gland, but is often inhibited by the neurosecretory MIH.

Therefore, the internal or external environmental conditions which are able to influence the molting pattern may, by way of the central nervous system, affect either the production of the MIH in the x-organ, or its release from the sinus gland. In this way the y-organs would be inhibited by a high blood titer of MIH, or would be activated to release MH by a low titer of MIH.

In some adult crustaceans molting continues until death (*Homerus,*

Cancer), while in others (*Maia, Carcinus, Pachygrapsus, Callinectes*) a final molt occurs after which the adult may live for an extended period of time. In the latter group the y-organ degenerates after the last molt, which further supports the suggested relationship of this gland to molting.

Similar observations have been made on the paired molting glands of insects, for they also degenerate following the final molt to an adult. It has therefore been suggested by many observers that the y-organs are not only analogous to the prothoracic glands of insects, but homologous as well. This homology is also supported by histological similarities.

Additional evidence which supports this suggested homology comes from the massive crayfish extractions and chemical analyses of two steroid molting hormones by Hampshire, Horn, and associates, in Australia (Hampshire and Horn, 1966; Galbraith *et al.*, 1968). Structurally, the two hormones are similar to that structure suggested for the insect molting hormone, ecdysone (Fig. 10-18). Crustecdysone

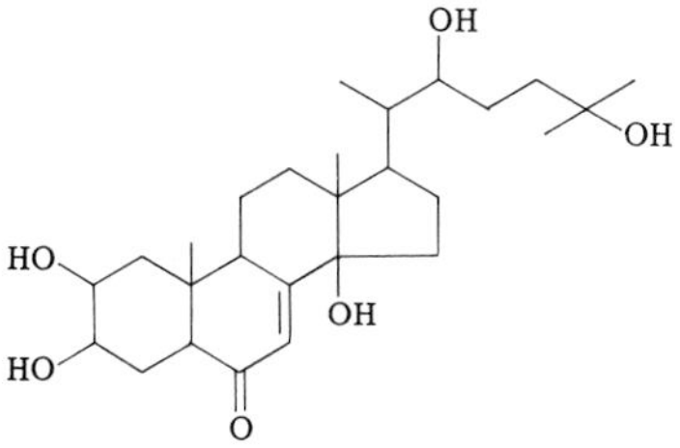

Ecdysone 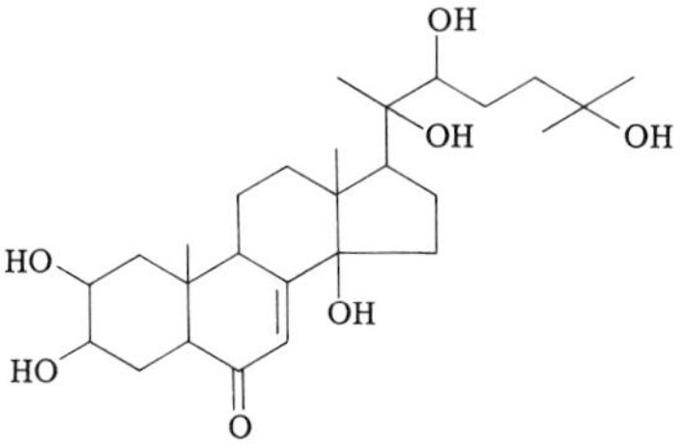   Crustecdysone

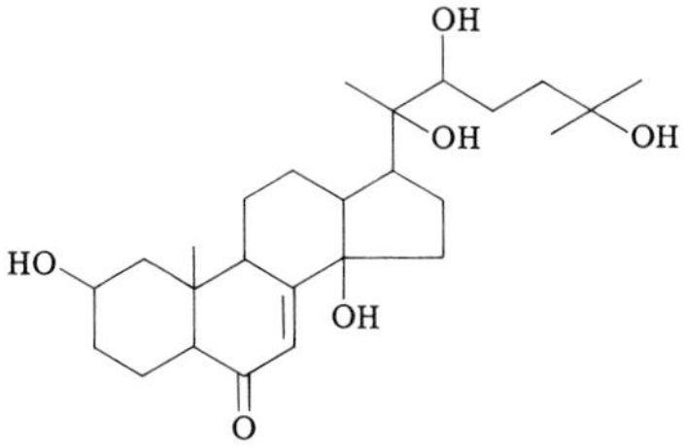

2-Deoxycrustecdysone

Fig. 10-18. Chemical structures for the insect molting hormone, ecdysone, and the two crustacean molting hormones, crustecdysone and 2-deoxycrustecdysone. (From Hampshire and Horn, 1966, and Galbriath *et al.*, 1968.)

and deoxycrustecdysone are the first two crustacean hormones which have been isolated and purified; although the substantially pure milligram quantities have yet to be crystalized, this isolation does represent a truly significant achievement in comparative biochemistry. This expansion of invertebrate endocrinology illustrates that chemical characterization of the invertebrate hormones is not limited solely to Insecta.

In light of the apparent homology for the molting gland on the part of malacostracous crustaceans and pterygote insects, it is worth mentioning that the activity of this gland is dependent upon opposite mechanisms. Activation in the y-organ depends upon the cessation of inhibition, whereas in insects molting gland activation is due to active stimulation. The inhibitor in crustaceans and the stimulant in insects are both neurosecretions; thus when the homologous neurosecretory perikaryons are removed from both, the crustaceans hasten molting, while molting is suppressed in insects.

A note concerning the partial characterization and isolation of the MIH appeared in 1965 and indicated that the neurosecretion appeared to be a peptide (Rangarao, 1965). Therefore its chemistry may be similar to the previously discussed eyestalk secretions: two chromatophorotropins and a distal retinal pigment hormone (Table 10-3).

5. REPRODUCTION

Hormones controlling sexual differentiation are produced by two nonneural structures: the androgenic gland in the male and the ovary in the female. Regulation and integration of the reproductive cycles, on the contrary, are more like the pattern observed in other invertebrates where neurosecretion influences the development of the gonads.

a. Sexual Differentiation

The majority of higher crustaceans are bisexual, with their sex genetically determined. At the time of hatching, their gonads and genital apparatus are undifferentiated, and several molts may pass before signs of sexual dimorphism develop. In the male, the control for this development is hormonal and is vested in the androgenic gland located on the vas deferens. This gland was discovered by Charniaux-Cotton (1954) in the amphipod *Orchestia gammarella* and, because of her excellent work on this species during the succeeding years, the endocrinology of reproduction of this species is best understood of all Crustacea. In most young, sexually undifferentiated

Crustacea a pair of rudimentary androgenic glands are present. In genetic females this structure fails to develop further, but in males the rudiment develops and all male sexual characteristics are dependent upon the secretions from this gland. If the androgenic glands alone are removed, the male reverts to a nonsexual or indeterminate stage at the next molt, while there is no effect following the extirpation of the testes or a portion of the vas deferens. If the androgenic glands are transplanted into a female, the ovaries and secondary female characteristics are masculinized in the following way: The primary germ cells give rise to spermatocytes, spermatids, and fertile spermatozoa, while the female appendages gradually acquire the male form. When an ovary is implanted into an operated male without androgenic glands it survives without modification but rapidly transforms into a testis when the host male possesses its androgenic glands, even though its testes may have been removed. This clearly shows the strong influence the androgenic gland possesses over differentiation of the male and indicates that the ovary is subordinate to its control.

Since the androgenic hormone is secreted into the blood, transfusion of the male hemolymph as well as an injection of an aqueous extract of the glands causes masculinization in recipient females.

The female does not possess a gland similar to the male's; however, there is evidence that in the female *Orchestia* one or two hormones are secreted by the ovaries. This secretion controls the appearance of several secondary sex characteristics: the oostegites and the ovigerous hairs on the oostegites. If the ovaries are removed, there is a loss of hairs which reappear in castrated females following reimplantation of ovarian tissue.

Unfortunately no information exists as to the chemistry of the male and female sex hormones. This information would be interesting since the crustacean's dependence upon hormones for sex differentiation is similar to conditions in vertebrates. One should not, however, reach too far to seek resemblances between two such unrelated groups and, indeed, there is no parallel within vertebrates of a gland entirely separate from the gonad yet capable of determining the primary and secondary sexual characteristics. There is also no similarity on this point between the crustaceans and insects, for sex hormones appear to play no role in sex differentiation of insects.

The phenomenon of parasitic castration which attracted much research attention several decades ago is simply due to damage to the androgenic gland by the parasite (a rhizocephalan barnacle, or epicaridian isopod). The result was the loss of the androgenic hormone

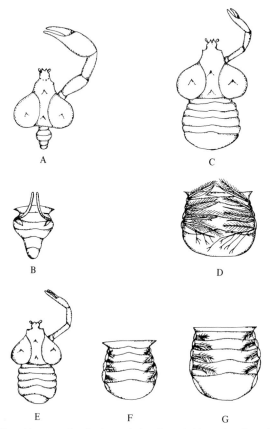

Fig. 10-19. Parasitic castration in the crab *Inachus*. A. Normal male with short small abdomen and broad claw; B. male abdomen without pleopods from beneath; C. normal female with broad abdomen and small claw; D. female abdomen from beneath with hairy pleopods; E. sacculinized male with female characters (small claw and broad abdomen); F. abdomen of sacculinized male from beneath with pleopods; G. abdomen of sacculinized female. (From Barrington, 1967, after Hanström, 1939.)

with the host reverting to a nonsexual condition or, in some cases, a feminine condition. The latter could be so extreme, with the testes developing as ovaries and brooding characteristics appearing, that sex reversal could be claimed for the host (Fig. 10-19).

Sex reversal is a natural condition in some crustaceans, and the biology of several proterandric hermaphrodites has been carefully studied. As with parasitic castration, the control of this form of reproductive development can be easily understood. The androgenic gland is initially present in the mature organism, which, therefore, promotes the production of spermatids during the male phase. At the appropriate time two changes occur in the hormone balance: The androgenic gland degenerates and the titer of the eyestalk inhibitory gonadotropin is reduced; together these signal the beginning of the female phase. At least one molt is required to obtain the necessary female body openings and appendages; however, internal changes may be more protracted before vitellogenesis can begin to be followed by egg laying.

b. Reproductive Cycles

In the normal development of a crustacean the gonads are not as independent from all other hormones as has just been indicated. There appears to be a secretion from the juvenile y-organ, possibly one of the molting hormones, which is also needed for the further differentiation of both the ovary and testis. When both y-organs are surgically removed in very young crabs, mitotic processes in the gonads of both sexes are impaired. In the ovary, oogonial mitosis stops, no follicles are formed around the oocytes, and without the follicles vitellogenesis does not occur. In the testis, spermatogonial mitosis ceases and the mature germ cells are therefore not formed. Thus in juveniles without the y-organ the gonads remain as they are, male or female, without any digression, but further development is decidedly limited.

Similar y-organ extirpations on adults either have no effect or stimulate the gonads. Demeusy (1962) found that the removal of the y-organ in *Carcinus maenas* favored ovarian enlargement. Charniaux–Cotton (1954) interpreted this as indicating that in the absence of the y-organ hormones only the nonmitotic processes could proceed. This would be the condition in the more mature organism with subsequent ovarian enlargement, but would not be possible in the immature where mitotic activity is occurring.

Except for the early development of the androgenic gland and gonads, most of the reproductive development is seasonal. This places the control of reproductive cycles under the neuroendocrine system, influenced by environmental stimuli which impinge upon various sensory receptors.

Initial experiments by Panouse (1943) indicated that when the eyestalks were removed from female *Palaemon serratus* in sexual rest, there was a rapid development of the ovary which led to early egg deposition. Thus, it is apparent that an eyestalk hormone acts to inhibit the last phase of oogenesis, which is probably the deposition of yolk. Selected removal of the x-organ and the sinus gland and their implantation into females preparing for reproduction have verified these observations.

This experimental procedure is not without its limitations, for when destalking occurs in immature females, the ovaries scarcely develop more than the controls. When eyestalks are removed from mature females shortly after a normal period of egg laying, no development takes place. In both cases the available oocytes probably had not reached the stage where yolk deposition could occur, and even though the hormones inhibiting yolk deposition has been removed the expected results were not observed.

Since there is both a molt-inhibiting and ovary-inhibiting (inhibitory gonadotropin) hormone produced by the eyestalk, a valid question is: How does the animal determine which of the two phenomena, both being somewhat antagonistic to one another, should be expressed following eyestalk removal? The answer seems to depend upon the time of the year, or annual cycle, during which the destalking occurs, since the animal is more disposed physiologically at one time for molting and at another time for reproductive development (Bliss, 1966).

Hormonal control of reproduction in Crustacea is still far from being completely understood, for there are now indications that an ovary-stimulating hormone may reside within the brain or thoracic ganglion capable of overriding the eyestalk inhibitor.

6. METABOLISM

A number of physiological systems show considerable change in activity at the time of eyestalk removal. These are comparable to changes associated with the two phenomena of molting and repro-duction in the intact animal. Such adjustments are indeed necessary for the satisfactory initiation and completion of the two phenomena. For that reason there is reluctance on the part of some investigators to consider these physiological systems as responding to direct hormonal influence, but, rather, they consider them to be indirect responses of a target organ system which is under hormonal control.

Oxygen consumption, calcium deposition, water balance, and blood sugar levels are four systems which can be easily studied and quantified. As mentioned in the section on molting, all four are known to change when the concentration of the eyestalk molt-inhibiting hormone declines, either as the organism is entering the normal premolt stage, or following experimental extirpation of the eyestalk. Studies concerned with oxygen uptake and calcium deposition during the premolt period have offered much to our understanding of molting, but evidence linking these systems to specific hormones is insufficient to justify any further consideration at this time. There is, however, stronger evidence that osmoregulation and the regulation of blood sugar are linked to specific hormones; in fact, a hypoglycemic factor has been partially characterized.

a. Osmoregulation

It has been suggested that the crayfish *Procambarus clarkii* is able to maintain normal body permeability, and thereby regulate body water, through the continued release of an eyestalk neurosecretion. When the organism approaches preecdyses, the concentration of this secretion in the blood is reduced, followed by an influx of water and a dilution of the blood concentration. This condition has been duplicated by the removal of the eyestalk or the sinus gland and, similarly, implantation therapy has returned the animal to a normal water balance. Kamemoto, Kato, and Tucker (1966) suggested that the maintenance of homeostasis in salt and water balance is a function of the neuroendocrine system, probably through the secretion of a neurosecretory factor.

Bliss, Wang, and Martinez (1966), working with the land crab *Gecarcinus lateralis*, suggested a somewhat different mechanism. They found no evidence for a hormone capable of inhibiting the uptake of water, but suggested, rather, that an antidiuretic hormone is released in the proecdysial period, thus reducing water loss and consequently facilitating water retention. The postecdysial period is accompanied by the secretion of a diuretic hormone which causes a loss of the retained water and, consequently, a return to the normal body concentration.

Both hypotheses may be correct for the organism studied, but little can be stated concerning a generalized scheme for Crustacea until more evidence is available.

b. Blood sugar regulation

More uniformity of opinion appears in reference to a possible hypoglycemic hormone, its origin, method of action, and characteristics, than in reference to any other endocrine factor influencing crustacean metabolism. This hormone was first named a diabetogenic factor by Abramowitz, Hisaw, and Papandrea (1944) because an elevation in the blood glucose concentration was obtained by the injection of eyestalk extracts. Conversely when eyestalks are removed, the blood sugar concentration decreases with an increase in the glycogen content of the hypodermis. This again is similar to the normal change that occurs in preparation for ecdyses when the molt-inhibiting hormone concentration is reduced.

From 1000 eyestalks of the prawn *Pandalus borealis,* Klienholz, Kimball, and McGarvey (1967) have separated and initially character-ized a fraction which they refer to as a hyperglycemic hormone (HGH) because it causes an increase in the blood glucose within 2 hours of the injection. Thus far the HGH appears to be a protein rather than a peptide.

Table 10-3 shows the chemical properties of five extractable eye-stalk hormones (Kleinholz, 1966). Four are believed to be peptides and one a protein. This observation takes on added significance when it is recalled that optical and electron microscopic observations showed five or six different types of neurosecretory granules in the axons of the sinus gland (pages 123 and 124).

References

Abramowitz, A. A., Hisaw, F. L., and Papandrea, D. N. (1944). The occurrence of a diabetogenic factor in the eyestalks of crustaceans. Biol. Bull. **86,** 1–5.
Barrington, E. J. W. (1964). Hormones and the control of color. In "The Hormones" (G. Pincus, K. V. Thimann, and E. B. Astwood, eds.), Vol. 4, pp. 299–363. Academic Press, New York.
Barrington, E. J. W. (1967). "Invertebrate structure and Function." Houghton, Boston, Massachusetts.
Belamarich, F. A., and Terwilliger, R. C. (1966). Isolation and identification of cardio-excitor hormone from the pericardial organs of Cancer borealis. Am. Zoologist **6,** 101–106.
Bern, H. A., and Hagadorn, I. R. (1965). Neurosecretion. In "Structure and Function in the Nervous System of Invertebrates" (T. H. Bullock and G. A. Horridge, eds.), Vol. 1, pp. 353–429. Freeman, San Francisco, California.
Bliss, D. E. (1966). Relation between reproduction and growth in decapod crustaceans. Am. Zoologist **6,** 231–233.
Bliss, D. E., Wang, S. M. E., and Martinez, E. A. (1966). Water balance in the land crab, Gecarcinus lateralis, during the intermolt cycle. Am. Zoologist **6,** 197–212.
Bunt, A. H., and Ashby, E. A. (1967). Ultrastructure of the sinus gland of the crayfish, Procambarus clarkii. Gen. Comp. Endocrinol. **9,** 334–342.
Carlisle, D. B., and Knowles, F. G. W. (1959). "Endocrine Control in Crustaceans." Cambridge Univ. Press, New York.
Charniaux–Cotton, H. (1954). Découverte chez un crustacé amphipode (Orchestia gammarella) d'une glande endocrine résponsable de la différénciation de caràctéres sexuels primaires et sécondaires mâles. Compt. Rend. **239,** 780–782.
Charniaux–Cotton, H., and Kleinholz, L. H. (1964). Hormones in invertebrates other than insects. In "The Hormones" (G. Pincus, K. V. Thimann, and E. B. Astwood, eds.), Vol. 4, pp. 135–198. Academic Press, New York.

Cooke, I. M. (1964). Electrical activity and release of neurosecretory material in crab pericardial organs. *Comp. Biochem. Physiol.* **13,** 353–366.

Cooke, I. M. (1966). The sites of action of pericardial organ extract and 5-hydroxytryptamine in the decapod crustacean heart. *Am. Zoologist* **6,** 107–121.

Cooke, I. M. (1967). Correlation of propagated action potentials and release of neurosecretory material in a neurohemal organ. *In* "Invertebrate Nervous Systems" (C. A. G. Wiersma, ed.), pp. 125–130. Univ. of Chicago Press, Chicago, Illinois.

Costlow, J. D., Jr. (1968). Metamorphosis in Crustaceans. *In* "Metamorphosis" (W. Etkin and L. I. Gilbert, eds.), pp. 3–41 Appleton, New York.

Demeusy, N. (1962). Rôle de la glande de mue dans l'évolution ovarienne du crabe *Carcinus meanas* Linne. *Cashiers Biol. Marine* **3,** 37–56.

Durand, J. B. (1956). Neurosecretory cell types and their secretory activity in the crayfish. *Biol. Bull.* **III,** 62–76.

Fingerman, M. (1965). Chromatophore. *Physiol. Rev.* **45,** 296–339.

Gabe, M. (1966). "Neurosecretion." Pergamon Press, Oxford.

Galbraith, M. N., Horn, D. H. S., and Middleton, E. J. (1968). Structure of deoxycrustecdysone, a second crustacean moulting hormone. *Chem. Commun.* No. 2, 83–85.

Gorbman, A., and Bern, H. A. (1962). "A Textbook of Comparative Endocrinology." Wiley, New York.

Hampshire, F., and Horn, D. H. S. (1966). Structure of Crustecdysone, a crustacean moulting hormone. *Chem. Commun.* No. 2, 37–38.

Hanström, B. (1939). "Hormones in Invertebrates." Oxford University Press, London and New York.

Hanström, B. (1947). The brain, the sense organs and the incretory organs of the head in Crustacea Malacostraca. *Lunds. Univ. Arsskr. N. F. Adv. 2.* **43,** 1–45.

Kamemoto, F. I., Kato, K. N., and Tucker, L. E. (1966). Neurosecretion and salt and water balance in the Annelida and Crustacea. *Am. Zoologist* **6,** 213–219.

Kleinholz, L. H. (1965). "Problems in Crustacean Endocrinology." Proceedings of the Symposium on Crustacea, Part III, pp. 1029–1037. Marine Biol. Assoc. of India.

Kleinholz, L. H. (1966). Separation and purification of crustacean eyestalk hormones. *Am Zoologist* **6,** 161–167.

Kleinholz, L. H., Kimball, F. and McGarvey, M. (1967). Initial characterization and separation of hyperglycemic (diabetogenic) hormone from the crustacean eyestalk. *Gen. Comp. Endocrinol.* **8,** 75–81.

Knowles, F. G. W. (1955). Crustacean colour change and neurosecretion. *Endeavour* **14,** 95–104.

Knowles, F. G. W. (1962). The ultrastructure of a crustacean neurohaemal organ. *In* "Neurosecretion" (H. Heller and R. B. Clark, eds.), pp. 71–88. Academic Press, New York.

Koller, G. (1928). Versuche über die inkretorischen Vorgange beim Garneelenfarbwechsel. *Z. Vergeleich. Physiol.* **8,** 601–612.

Matsumoto, K. (1958). Morphological studies on the neurosecretion in crabs. *Biol. J. Okayama Univ.* **4,** 103–176.

Matsumoto, K. (1959). *Biol. J. Okayama Univ.* **5,** 43 (cited by Gabe, 1966).

Panouse, J. B. (1943). Influence de l'ablation du pédoncule oculaire sur la croissance de l'ovaire chez la Crevette *Leander serratus*. *Compt. Rend.* **217,** 553–555.

Passano, L. M. (1961). The regulation of crustacean metamorphosis. *Am. Zoologist* **1,** 89–95.

Perkins, E. B. (1928). Color changes in crustaceans, especially in *Palaemonetes*. *J. Exptl. Zool.* **50,** 71–105.

Potter, D. D. (1958). Observations on the neurosecretory system of portunid crabs. *In* "Second International Symposium on Neurosecretion" (W. Bargmann *et al.*, eds.), pp. 113–118. Springer, Berlin.

Pouchet, G. (1876). Des changements de coloration sours l'influence des nerfs. *J. Anat. Physiol.* **12,** 1–90 and 113–165.

Rangarao, K. (1965). Isolation and partial characterization of the molt-inhibiting hormone of the crustacean eyestalk. *Experientia* **21,** 593–594.

Scharrer, E. (1958). General and phylogenetic interpretations of neuroendocrine interrelations. *In* "Comparative Endocrinology" (A. Gorbman, ed.), pp. 233–249. Wiley, New York.

Turner, C. D. (1966). "General Endocrinology." Saunders, Philadelphia, Pennsylvania.

CHAPTER 11

ARTHROPODA—MYRIAPODA, INSECTA

I. INTRODUCTION

The truly successful invertebrate animals, the mandibulate arthropods, have adapted to humid as well as arid habitats, and comprise between one-third and one-half of all animal species. In contrast to the aquatic mandibulates, the Crustacea with their second pair of antennae (antennules) and biramous appendages, the terrestrial mandibulates have no antennules and possess uniramous appendages. Two evolutionary paths have been taken by the terrestrial mandibulates, one leading to the dominant group—insects—characterized by the three-body tagmata of head, thorax, and abdomen, and the other leading to the lesser group—myriapods—containing only two tagmata, head, and trunk. In this introductory discussion of the endocrinology of these two groups, the myriapods will receive only a passing glance.

II. MYRIAPODS: DIPLOPODA AND CHILOPODA

The general pattern of the nervous system of arthropods which has been discussed for chelicerates and crustaceans is applicable also to the myriapods. The three components of a neurosecretory pathway are present in this group: the perikaryon for synthesis, the axon for transport, and a neurohemal organ for storage and release. Neurosecretory cells have been found in all areas of the central nervous system, with the most numerous groups located in the protocerebrum. Little difficulty is encountered in following the paths of the neuro-secretory axons from the cell body to the neurohemal organ. The cerebral gland is the principal storage structure in the myriapods and varies considerably in its appearance and location between groups. The gland may appear as a group of small lobes bunched together, or as a compact body; it may be located laterally to the brain and ventrally to the optic lobes, or more posterior and medial to the brain; it may be attached to the brain by a short nerve, thereby remaining close to the brain surface, or with a long nerve, resting against the ventral body wall. However, the gland always remains in the head region. As in other arthropods there appears to be intrinsic secretory activity within this neurohemal organ, and the product of the intrinsic cells does not appear to stain with paraldehyde-fuchsin as is the case with the neurosecretory axons entering the organ. Neither ultra-structural nor histochemical studies are available to provide further information on the nature of the intrinsic cell secretion (Fig. 11-1).

In the class Diplopoda, eleven genera have been studied histologi-cally and found to contain neurosecretory cells. These are primarily located in two symmetrical groups in the protocerebrum, but are also found in the tritocerebrum as well as the subesophageal and ventral nerve cord ganglia. The secretory cells of the protocerebrum are

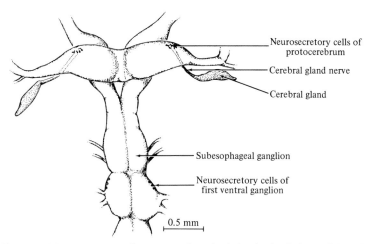

Fig. 11-1. Neurosecretory cell groups and cerebral glands of *Lithobius* (Chilopoda). (From Bern and Hagadorn, 1965, after Palm, 1955.)

piriform in shape, measuring from 20–25μ with nuclei centrally placed. Prabhu (1962) classified the cells of *Jonespeltis splendidus* into three types which could then be subdivided according to the appearance of the secretory product. Such material can be observed in the axons of neurosecretory cells which course through the cerebral ganglion and emerge at the boundary between the protocerebrum and deutocerebrum. At this point of emergence they form a pair of nerves which terminate in the bilateral cerebral glands, one of the two paired neurosecretory cells of the tritocerebrum, and possibly by subeso-the hypocerebral organs, or connective bodies, attached to the circumesophageal connectives and believed to be innervated by the neurosecretory cells of the tritocerebrum, and possibly the subeso-phageal cells.

No functional significance has yet been associated with any cells in Diplopoda except in the findings of Prabhu (1962), who recorded morphological changes in all three neurosecretory cell types of *Jonespeltis* during the course of the annual cycle.

Neurosecretion has been reported in twenty species representing nine genera of the class Chilopoda and, generally, the morphology is similar to that of the Diplopoda. Secreting cells are located in compact bilateral groups, usually on the dorsal aspect of the protocerebrum out of which axons lead ventrolaterally to form the cerebral gland nerve. Little functional evidence is available for the neurosecretory cells; however, R. Joly (1962) suggests that the cerebral glands are involved in centipede molting.

III. INSECTA

This group of invertebrates represents the culmination of the terrestrial mandibulate line. The class contains thirty-two orders and is characterized best by their three-body regions—head, thorax, and abdomen—with three pairs of legs and, with few exceptions, two pairs of wings attached to the middle body region, the thorax. Although the class is large with many adaptations, no taxonomic divisions will be made in this discussion. The primitive wingless Apterygota, consisting of four orders, present several interesting morphological and functional features, each of which will be discussed under the appropriate section.

Kopec, the Polish zoologist, could rightly be considered the father of insect, as well as invertebrate, endocrinology. Some 50 years ago (1917, 1922), when he provided the first evidence that the brain of a caterpillar is functionally important in pupation, he not only initiated work on invertebrate endocrinology but, more importantly, implicated nervous tissue in processes of chemical integration. During the 1930's V. B. Wigglesworth, at Cambridge University, was instrumental in linking morphology with experimental physiology, and thereby initiated a notable career in exploring the hormonal bases of insect phenomena. After the revolutionary concept of neurosecretion was introduced by the Scharrers in the early 1930's, the first example of neurosecretion in insects was described in *Apis mellifera*, the honey bee, by Weyer (1935). From this point on, research in morphology, physiology, and endocrinology of insects developed in a coordinated way so that today the insects are by far the most extensively studied, and consequently the best understood of all invertebrates. This state of affairs is reflected in the large number of general review articles on insect hormones; however, in recent years such reviews have become more limited to specific areas of hormone influence, such as growth and development, reproduction, diapause, and metabolism. The first book to appear on insect hormones was written by Novak in German some 10 years ago; its third edition appeared in English in 1966. A second book on insect endocrinology was published in French by P. Joly in 1968.

Thus, by virtue of the insect's worldwide distribution, economic importance, ease of handling and rearing, and the excellent beginning by the European zoologists, insect endocrinology is well understood compared to other invertebrate groups. For information more detailed than that presented in this chapter, the above two books and the reviews mentioned throughout the chapter should be consulted.

A. Morphology

The nervous system in insects is essentially the same as that for any mandibulate arthropod. Three further developments, however, can be noted: (1) continued development and specialization of specific nuclei and relay centers of the brain; (2) centralization through anterior migration and coalescence of the ventral nerve cord ganglia; and (3) necessary lengthening of the peripheral nerves to compensate for the forward movement of the ganglia. The insects are also similar to other arthropods in their neuroendocrine systems with the following three basic parts: (1) neurosecretory cells present within most if not all of the ganglia of the central nervous system; (2) a well-defined neuro-hemal structure posterior to the brain; and (3) two glands of ectodermal origin in the head or thorax (Fig. 11-2).

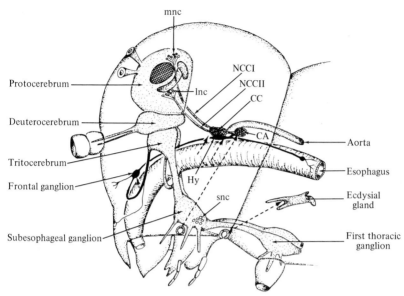

Fig. 11-2. The generalized central nervous system and principal endocrine glands of a hemimetabolous insect are presented in lateral view. The stomatogastric, or visceral, nervous system arises from the stomodaeal ectoderm, and has paired nerves from the tritocerebrum and the median frontal ganglion. A recurrent nerve extends along the esophagus to the hypocerebral ganglion (Hy). Hormones are secreted by median (mnc), lateral (lnc), and subesophageal neurosecretory cells (snc). Secretions from the cells of the protocerebrum pass in two paired nerves (NCCI and NCCII) to the corpora cardiaca (CC), which arise from stomodaeal ectoderm (dashed arrow). Hormones are also secreted by two paired glands that arise as ectodermal invaginations: the corpora allata (CA) which migrate to above the esophagus where they receive axons from the corpus cardiacum, and the ecdysial glands which move usually into the prothorax. (From Jenkin, 1962, after Weber, 1949.)

1. NEUROSECRETORY CELLS—CEREBRAL GANGLION

The protocephalic neurosecretory cells and their pathways were the first to be discovered and have since received more attention from the morphologists and physiologists than any of the other similar cell groups in the insect's central nervous system. There are two main cell groups in each hemisphere symmetrically placed in the dorsal and dorsolateral surface of the brain. Ultrastructural and histochemical studies have indicated strong similarities between this system and that of the neurosecretory cells in the supraoptic and paraventricular nuclei of the vertebrate hypothalamus. These similarities lead to the prediction, which so far has not been fully supported, that the insect neurosecretion would prove to be a polypeptide or protein, and thus also show chemical similarities (see page 169).

The bilateral cells situated near the median furrow in the pars intercerebralis are quite constant in location in all pterygote insects, and are labeled A-cells, or the medial cell group (Fig. 11-3). The number of cells and their dimensions, which vary from order to order, are often difficult to ascertain due to their fused appearance, but the cell count ranges from 10–65 in the honey bee worker, to 18 in the cecropia silkworm. The largest diameter of the cell body in the Odonata is 20 μ but may approach 70 μ in Lepidoptera and Hemiptera. Most cell bodies are piriform but are often greatly distorted by adjacent cell pressure. The processes from the medial cell group as they emerge from the brain form a pair of nerves, the nervi corporis cardiaci I (NCCI), which extend to and innervate the corpora cardiaca. The intraganglionic course of these axons is rather uniform among insects. From the dorsally positioned cell bodies, the axons extend rostrally, crossing to the opposite sides in the midline, then turn

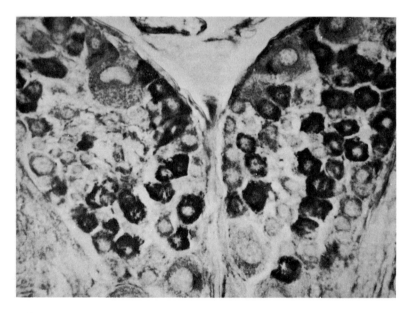

Fig. 11-3. Section of pars intercerebralis of protocerebrum from *Periplaneta americana* adult. Paired groups of medial neurosecretory cells showing large numbers of fuchsinophilic and a few large acidophilic cells. Axons within NCCI extend from these cells to the corpora cardiaca. (From Gorbman and Bern, 1962.)

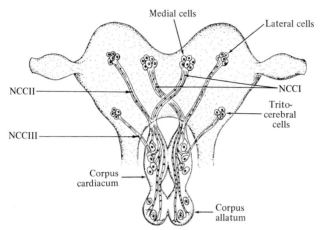

Fig. 11-4. Diagram of the neurosecretory system of insects showing the decussation of the NCCI and the tritocerebral neurosecretory cells, which have been noted in the phasmids. (After Knowles, 1963.)

aborally so that they exit the cerebral ganglion in a transverse plane, coming to lie adjacent to the aorta and stomodaeum (Fig. 11-4).

The second pair of cells are located on the dorsolateral surface of the brain, near the corpora pedunculata, and about halfway to the lateral edge of each cerebral lobe. These lateral or B-cells are usually smaller in size and almost always fewer in number than the medial cells. Cecropia has ten lateral cells, five on each side. Axons from these cells run a much simpler and more direct intraganglionic course from the perikaryon to the point of exit, there forming the paired nervi corporis cardiaci II (NCCII) which are lateral to the exit of the NCCI.

Although this is the generalized picture of neurosecretory cells in the cerebral ganglion of adult insects, a number of modifications and exceptions exist which will be illustrated by the following three examples. An additional pair of cells is located in the tritocerebrum of several phasmids which send axons directly to the corpora cardiaca via a third nerve trunk, the NCCIII (Fig. 11-4). The larval arrangement of neurosecretory cells may be considerably more complex, as Fraser (1959) has identified six bilateral groups of cells in the brain of the Diptera *Lucilia*. Among the apterygote insects, the protocerebral neurosecretory cells of the order Thysanura are anatomically separated from the brain in an encapsulated area which constitutes the frontal organs. Axons from cells in these organs form two nerve trunks which correspond to the paired NCCI of pterygote insects, and innervate the corpora cardiaca.

Ventral Nerve Cord

Practically all insect orders have demonstrated at least one pair of neurosecretory cells in the subesophageal ganglion. In this ganglion in *Leucophaea,* eight cells have been classified into three types on the bases of size, location, and staining characteristics. One pair has been termed the "castration cells" since an alteration in their cytology follows removal of the ovaries. The same ganglion in *Bombyx mori* has been reported to contain eighty cells!

The direction taken by the neurosecretory axons within, and as they emerge from, the subesophageal ganglion is not well known. In some cases axons cross to the opposite side and extend from the dorsal surface of the ganglion directly to the prothoracic glands, as in Odonata, or by way of the corpora allata, as in Ephemeroptera. However, in most cases the axon density in the ganglion is so light that complete resolution of the intraganglionic pathways and the morphological location of the axon bulbs will have to wait for improved staining techniques or ultrastructural examination (Fig. 11-5). Mor-

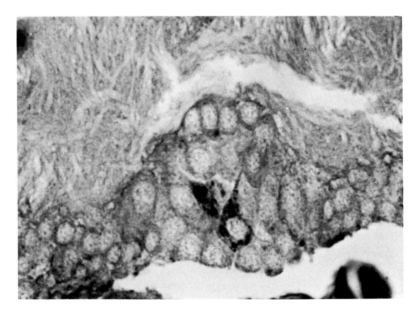

Fig. 11-5. A pair of neurosecretory cells in the subesophageal ganglion of an adult *Hypera postica.* × 900.

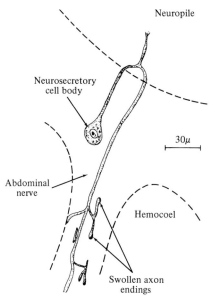

Fig. 11-6. A neurosecretory cell and its axon at the back of the mesothoracic gan-
glionic mass of *Rhodnius prolixus*, with sites of release within the abdominal nerve.
The dotted lines indicate the edges of the neuropile and of the ganglionic mass. (From
Maddrell, 1966.)

phologically, neurosecretory cells in the thoracic and abdominal
ganglia have been mentioned rather incidentally by those more concerned with other aspects of neurosecretion; thus, we are aware of
their presence in almost all ganglia, but know little more than that.
Herlant-Meewis, Naisse, and Nouton (1967) provide the most recent
review of neurosecretion within the invertebrate ventral nerve cord,
with a third of the article devoted to insects. In a review of neurosecretion Maddrell (1967) suggests that the ventral nerve cord is an
area which offers many challenging questions for the neuroendocri-
nologist. Not only is the morphology virtually unknown, but there
are very probably a number of hormones yet to be discovered which
originate in this region of the body (Fig. 11-6).

2. NEUROHEMAL ORGAN

The corpora cardiaca, principal insect neurohemal organs, are
small, paired structures, bluish in appearance when viewed *in situ,*
located just posterior to the brain, in close association with the
dorsal aorta and esophagus. They have already been described as

receiving one, two, or three nerves, i.e., NCCI, II, or III, from the brain. Each cardiacum is connected to its corresponding corpus allatum by a single nerve, nervus corporis allati, composed of axons originating in the brain which continue through the cardiacum into the allatum. The possibility also exists, however, that axons from cells within the cardiacum may also be incorporated into this neuron.

The cardiaca arise embryonically from cells of the stomodaeal wall, in a similar fashion to that in the origin of the dorsal sympathetic nervous system (Fig. 11-2). On this basis, the cardiaca could be considered as modified nerve ganglia, and if secretory cells exist in the body they could be considered as being neurosecretory, although at present this classification is debatable.

A large portion of the cardiacum is made up of the swollen axon terminals entering from the brain (Fig. 11-4). The neurosecretory product may accumulate there or be released immediately into the hemolymph (Figs. 2-4, 2-5). While the material is in the distal region of the axon, some chemical change may occur, activating the product before it passes across the membrane. In some apterygotes, the protocerebral neurosecretory axons end in the aortic wall rather than the cardiacum, facilitating the distribution of the active material via the hemolymph.

In addition to the neurohemal role of the cardiacum, an additional function has been suggested for this structure. On the basis of epithelioid elements which appear to be secretory in nature and undergo a cycle of activity, intrinsic secretory cells have been identified in cardiaca from many species. In Lepidoptera the secretory cells are few, but in Acrididae they form rather large areas somewhat separate from the neurohemal region. Thus, if the cardicum can act as an endocrine as well as a neurohemal organ, a number of questions can be posed which concern the nature of the secretory cells, their independence from or dependence upon the control of the brain, and the nature of the physiological system they influence.

Although no function can be associated with their presence, true nerve cells have been reported in the cardiacum of a few species. This is in accordance with the accepted ontogeny of the organ as mentioned earlier, and also supports the conclusion of earlier morphologists that the cardiacum is an ordinary ganglion.

Combined with the axon endings, intrinsic secretory cells, and nerve cells, the fourth component identified is connective tissue, or glia. The glial cells are wrapped around the axons within the NCC trunks, and clearly function in the same role within the cardiacum.

Although the cardiacum is without question the principal insect neurohemal organ, it is not alone in fulfilling this function. The nerve connecting the cardiacum with the allatum carries with it a large number of neurosecretory axons from both the medial and lateral neurosecretory cells of the brain; these then release their contents into the cells of the allatum, or into the hemolymph. Other neurosecretory axons have been traced from the brain into the subesophageal ganglion. An extensive system has been observed in aphids where neurosecretory axons extend almost the length of the body to directly innervate specific target organs. With the distribution of neurosecretory cells now including almost all ganglia of the ventral nerve cord, neurosecretory storage and release areas have been and will continue to be found on the surface of the ventral nerve cord or peripheral nerves (Fig. 11-6).

In summary, the cardiacum is no longer a simple storage and release structure for the products of the protocerebral neurosecretory cells, but is a rather complex organ with endocrine and nerve cells, as well as the axon endings. Nevertheless, the cardiacum remains the principal, but no longer the only, area for the release of neurosecretory material in insects.

3. NONNEURAL ORGANS—CORPORA ALLATA

Extending from the distal region of the cardiaca in most insects, except Thysanura and most Ephemeroptera, are the nervi corporis allati which innervate the corpora allata. Usually bilaterally arranged, the allata have no ontogenetic relationship with the cardiaca, or any neurosecretory pathways. The allata originate from ectodermal invaginations between the mandible and first maxillary segments. The paired cluster of cells move dorsally and posteriorly into the center of the head capsule near the esophagus and aorta, and there receive the pair of nerves from the cardiaca (Fig. 11-2). These connecting nerves are primarily extensions of the nervi corporis cardiaci; thus, the allata become a segment of the protocephalic neurosecretory pathway which extends from the protocerebrum through the cardiaca into the allata. The presence of neurosecretory axons in the allata has been observed, after appropriate staining, with both optical and electron microscopy (Fig. 11-7). These axons with varying concentrations of elementary neurosecretory granules extend and then terminate both within the body of the allata, where contact with the bulk of glandular cells might easily occur, and around the periphery of the gland,

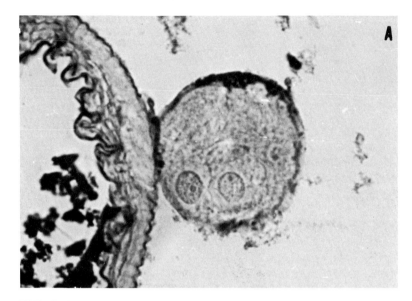

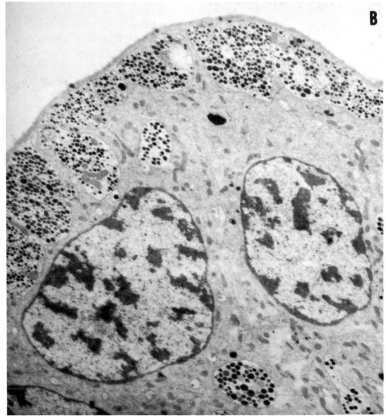

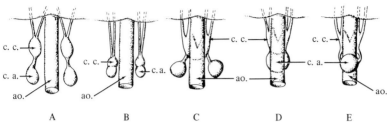

A   B   C   D   E

Fig. 11-8. The five major arrangements of the corpora cardiaca (c.c.) and corpora allata (c.a.) in relationship to the brain and aorta (ao.). Each type is discussed in the text. (From Bern and Hagadorn, 1965, after Cazal, 1948.)

which would facilitate release of the neurosecretory material to the hemolymph.

The unit arrangement of the cardiaca and allata, which form the retrocerebral complex, can be classified into five types as illustrated in Fig. 11-8: (A) simple bilaterally symmetric (Trichoptera, Lepidoptera, Homoptera); (B) distal bilaterally symmetric with corpora cardiaca allata fused (some Diptera and some Coleoptera); (C) semicentralized, with the corpora cardiaca partly fused (most apterygotes, Mallophaga, Orthoptera, Blattodea, Mecoptera, Hymenoptera, some Diptera, and some Coleoptera); (D) centralized with fused corpora cardiaca and allata together in one body (Pleocoptera, Dermaptera, Psocoptera, many Homoptera and Heteroptera, and brachyceran Diptera); (E) fused ring type forming, around the aorta, the Weismann's ring of higher Diptera (Bern and Hagadorn, 1965). The single ventral cardiacum in Diptera is connected to the single dorsal allatum by a pair of nervi corporis allati which are embedded in the ecdysial gland, forming a composite endocrine structure (Fig. 11-9).

The corpus allatum is usually solid and ovoid in shape but two marked exceptions can be found. In some Lepidoptera, a loose cluster of cells is found, resembling a cluster of grapes, while the phasmids possess a hollow gland with a well-defined central cavity. The cells of all corpora allata present a homogeneous picture cytologically, which is in sharp contrast to that observed in the cardiaca. In the inactive

---

Fig. 11-7. Corpora allata of two weevils. A. An allatum of *Anthonomus grandis* in cross section just lateral to the gut, with fuchsinophilic material concentrated around the periphery of the gland. × 900; B. An electron micrograph of an allatum from *Hypera postica*. Axons containing elementary neurosecretory granules are concentrated along the periphery of the gland, but are also located internally. × 8,550.

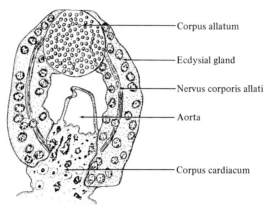

Corpus allatum

Ecdysial gland

Nervus corporis allati

Aorta

Corpus cardiacum

Fig. 11-9. A section through the retrocerebral complex (ring gland, Weismann's ring) of a young pupa of *Eristalis* (Diptera). (From Bern and Hagadorn, 1965, after Cazal, 1948.)

state, the nuclei are densely packed with reduced cytoplasm and considerable interdigitation of cell membranes. When active, the cytoplasmic content increases, cell membranes are straightened, and an increase is observed in mitochondria and endoplasmic reticulum. From similar ultrastructural evidence the suggestion has been made that the allatum resembles vertebrate glandular tissue which is active in steroid biosynthesis; therefore, the secretions of this gland would also be steroid in nature.

Ecdysial Gland

The second nonneural endocrine gland of insects arises in the embryo from ectodermal cells at the base of the second maxilla, and assumes a position which is less certain than that of the allata (Fig. 11-2). In the larvae of cyclorrhophous Diptera, the gland becomes associated with the retrocerebral system (Fig. 11-9), but in all other orders the paired structure is found either in the head or thorax. In the Thysanura, Ephemeroptera, Odonata, Isoptera, Dermaptera, and some Orthoptera, the glands remain compact and in the lower part of the head. In other Orthoptera, Hemiptera, Coleoptera, Lepidoptera, and Hymenoptera, the position of the gland is in the thorax and the form of the gland is considerably less compact; in fact, in Lepidoptera it is considered to be diffused. The conclusion is easily made that the more advanced the insect, the more diffused is the gland, and the more posterior will be its location (Fig. 11-10). Because of this variation in the position of the gland, many names have been assigned to this structure by morphologists, understandably causing some difficulty in interpreting the literature. The accepted term now is ecdysial gland, which emphasizes its functional importance in the control of molting (Herman, 1967). Herman's article is the only comprehensive review devoted to the cytology of the ecdysial glands of insects and should be consulted for further information on this subject.

The glands, whether in the head or thorax, are usually associated with the tracheal system and are made up of a homogenous type of secretory cells, including a few hemocytes, tracheal, epithelial, and glial cells, as well as scattered muscle fibers. Nervous innervation in certain groups originates in the subesophageal or thoracic ganglion, but the glands of a number of orders are apparently without innervation. Activation of the gland apparently is dependent upon the release into the hemolymph of an ecdysiotropic hormone from the neuro-

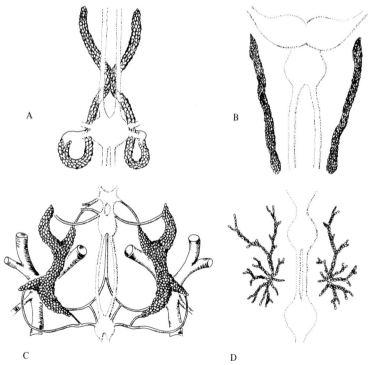

A   B   C   D

Fig. 11-10. The ecdysial glands in four insect orders: A. Orthoptera; B. Hemiptera; C. Lepidoptera; D. Hymenoptera. (From Novak, 1966, after various authors.)

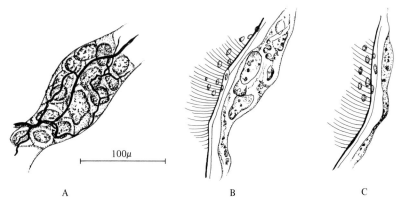

A   B   C

Fig. 11-11. Ecdysial gland of the insect *Dysdercus* showing degeneration following adult emergence. A. Active gland; B. 24 hours after final molt; C. 36 hours after final molt. (From Gorbman and Bern, 1962.)

secretory axons in the corpora cardiaca. Recently, electron microscopic examinations have supported earlier observations that stainable inclusions were present in nerves, innervating the ecdysial glands by revealing axons containing elementary neurosecretory granules within ecdysial glands of cockroaches, moths, and flies. Two distinct types of granules have been observed, suggesting that two separate responses may be under neuroendocrine control. Therefore, the control of the gland could be more complex than has been earlier assumed, with both the synthesis and release under separate hormonal control (Herman, 1968).

As in the corpora allata, cycles are observed in the secretory activity of the prothoracic glands which parallel the molting cycle. In most cases the maximum size of the gland is reached just prior to ecdysis, with a gradual reduction occuring after the molting process has been initiated. This cycling occurs for each molt throughout the insect's life, with glands degenerating after the adult emerges (Fig. 11-11).

Correlations are thought to exist between the extent of the agranular endoplasmic reticulum development and the peak period of hormone synthesis. This is supported by evidence that the molting hormone is a steroid (See page 170).

B. The Chemical Nature of Insect Hormones

Encouraged by the possibility that insect pests may be managed by the use of synthetic insect hormones or their analogues, insect endocrinologists and organic chemists have, in the last few years, joined forces in this area of research and have made several noteworthy discoveries.

The success which the vertebrate endocrinologists have had in modifying physiological processes with synthetic hormones should strengthen the research effort and provide further encouragement for those interested in the identification of insect as well as other invertebrate hormones. An indication of what may be rightly referred to as the "third generation insecticides" is the reported success obtained in the use of synthesized cecropia juvenile hormone in limited trials on coddling moth infestation of apple trees (Williams and Robbins, 1968).

The three hormones which have received the attention of the chemists are: (1) the ecdysiotropin, synthesized by neurosecretory cells of the protocerebrum and released to the hemolymph by the corpora cardiaca; (2) ecdysone, the secretion of the ecdysial gland; and (3) juvenile hormone, synthesized within and released from the corpora allata.

1. ECDYSIOTROPIN (BRAIN HORMONE, ACTIVATION HORMONE, PROTHORACOTROPIN)

Based on ultrastructural and histochemical evidence, it was predicted that the active neurosecretory material would be similar to that of vertebrates, i.e., a small polypeptide or protein. The first report of a successful attempt to find an active brain substance in insects was made by two Tokyo scientists, Kobayashi and Kirimura (1958). They removed and extracted 8500 brains from pupae of *Bombyx mori* and, to their surprise, the active fraction had no proteinaceous properties but was soluble in both methanol and ether, with physical properties similar to those of cholesterol. Their test of activity was the induction of pupation in brainless *Bombyx* pupae following quantitative injection of an unknown fraction. This indicated rather clearly that some inconsistencies existed between the experimental results and those predicted for the chemistry of the hormone. This was further complicated by the report of a second Japanese team, Ichikawa and Ishizaki (1961), that the brain hormone, also extracted from *Bombyx* but tested on brainless *Samia cynthia ricini,* was a water-soluble, heat-stable, nondialyzable substance which, they concluded, must be a polypeptide or protein. In 1962 Kobayashi and associates reported further evidence substantiating their hypothesis that the brain hormone was cholesterol. This evidence was based on chemical characterization of 4 mg of crystals prepared from 220,000 brains dissected from *Bombyx* pupae. Evidence continued to appear in the literature supporting both views, but without any clear reconciliation between the two until, in 1966, Kobayashi and Yamazaki partially remitted by confirming the presence of proteinaceous properties in the active fraction, though not completely abandoning their earlier conclusion that cholesterol was also a component. This view was further strengthened in 1967 with the report by Ishizaki and Ichikawa of an 8000-fold purification of the active principle, with the molecular weight of the major components from 9000–31,000.

In 1967 Williams summarized this entire topic quite lucidly as well as adding experimental observations from his research which supported the water-soluble, heat-stable, and nondialyzable contentions of others. Williams, however, did not extend his support to the suggestion that the hormone is a protein, on the basis that the hormone exhibits resistance to several peptidases. Thus the chemical nature of the hormone is still debatable.

2. ECDYSONE (MOLTING HORMONE, PROTHORACIC GLAND HORMONE, GROWTH AND DIFFERENTIATION HORMONE)

Stimulation of ecdysial glands by the ecdysiotropin brings about the release of a hormone which initiates molting. Work was begun in the late 1930's on the chemistry of this molting hormone but the first significant finding was that of Butenandt and Karlson (1954) when they isolated 25 mg of crystalline material from 500 kg of silkworm pupae. This was the first hormone isolated from any invertebrate. From this work the accepted name of the hormone emerged as well as the establishment of a test which is still used in isolation and purification procedures for the hormone. This test utilizes the nonpupated, posterior two-thirds of a ligated *Calliphora* prepupa into which an unknown fraction is injected. If between 50–70% of the injected individuals pupate in 24 hours, then that quantity of hormone injected is considered as "one *Calliphora* unit." It has now been determined that this unit is equal to approximately 0.0075 μg of α-ecdysone.

The Karlson group in Germany has continued research on ecdysone and, through them, our concept of the chemical nature of this hormone has continued to improve. Starting with four tons of fresh silkworm pupae, an equivalent of 1000 kg of dry weight, Karlson and his associates were able to obtain 250 mg of crystalline ecdysone. From this material sufficient analyses were conducted so that in 1963 the hormone was definitely reported as a steroid, with a molecular weight of 464, and with an empirical formula of $C_{27}H_{44}O_6$ (Karlson and Hoffmeister, 1963). Three years later it was discovered that the active fraction was composed of two steroids, α-ecdysone and its 20-OH derivative, β-ecdysone. The derivative is present in smaller quantities than the α form but is more active biologically. The derivative also appears to be identical to crustecdysone discussed in the previous chapter (Fig. 10-18).

A number of ecdysone-like materials, i.e., phytoecdysones, as well as α- and β-ecdysones, have recently been extracted from evergreen trees, weeds, and ferns in America, Europe, and Asia, and the number of similar compounds certainly will continue to increase. This research activity is encouraged by the finding that some analogues may, in certain insects, inhibit larval growth and development, or may inhibit adult ovarian development and reproduction. In this way the plants possess a natural protection against insect attack through the evolution of chemicals which, although very close in chemical structure to the molting hormones, are inhibitory to normal insect growth and development (Robbins et al., 1968).

The future for using synthetic hormones or their analogs in insect management is indeed bright, and the ecdysones appear to offer an excellent family of candidate compounds.

3. JUVENILE HORMONE (NEOTENIN, CORPUS ALLATUM HORMONE)

The hormonal secretion of the corpora allata, which appears not only in the immature stage but also in adults, was extracted initially from the abdomen of adult male Hyalophora cecropia by Williams (1956a). Two-day-old males of this species have, for some unexplained reason, the largest titer of hormone by far of any stage in the developing moth, which is also thirty times the level found in females of similar age. The active principle is obtained, when cecropia are extracted with cold ethyl ether, as a heat-stable, nonsterol, unsaponifiable lipid. A number of tests are used in evaluating the fractions for juvenile hormone activity. All bioassays basically involve the injection or application of the suspected material and an oil base into or onto a pupa at the beginning of adult differentiation and the examination of the adult form for the retention of larval characteristics (Fig. 11-12).

For 8 years after Williams' initial extraction, a number of laboratories attempted to characterize the hormone, but to no avail. The progress made and the experimental approach taken by these investigators during that time are presented in a review by Gilbert (1964).

Positive responses have been obtained in juvenile hormone bio-

Fig. 11-12. Effects of synthetic juvenile hormone on growth and development in two insects. A. The normal mealworm pupa (center) responds to the chemical by molting to a pupal adult intermediate (left), or undergoing an additional molt to form a second pupa (right); B. a similar application to the last instar nymph of the milkweed bug results in the formation of a supernumerary nymph (center) at the next molt, rather than a normal adult (right). (From Bowers, 1968.)

172

assays with extracts obtained from a rather wide variety of materials, i.e., from vertebrate and other invertebrate tissue, as well as from plant tissue. The very interesting "paper factor" was reported by Slama and Williams (1965), and showed that a material, chemically similar to the juvenile hormone, was present in some finished paper products which was at a concentration high enough to prevent metamorphosis in the sawfly *Pyrrhocoris*. This substance, now isolated and identified from balsa fir, called "juvabione" (with a related compound, "dehydro-juvabione") differs from other extracts in that it is a monocyclic sesquiterpene and is selectively active on the family Pyrrhocoridae. (Bowers *et al.*, 1966; Cerny *et al.*, 1967). The active material was originally thought to be located only in American coniferous trees, but it is now believed to be only one of many substances more universally located in plant tissue.

By 1967 all indications pointed to the triple isoprene unit of farnesol as providing the basic structure of the hormone, with a methyl ester derivation being one of the most active. The structure of the cecropia silkworm hormone was finally determined by Röller *et al.* (1967), and proved to have very definite affinities with farnesol (Fig. 11-13). This substance possesses all the gonadotropic and morphogenetic activities associated with the corpora allata and is active at sub-microgram levels when assayed on representatives of the Coleoptera, Lepidoptera, Hemiptera, and Orthoptera.

Many analogues are now being synthesized which seem to possess equal or greater biological activity than that of the isolated cecropia hormone. W. S. Bowers *et al.* (1965) synthesized 10,11-epoxy-farnesenic acid methyl ester (Fig. 11-13) which demonstrates considerable juvenile hormone activity in bioassay tests and which terminates adult diapause in representatives from several insect orders. Slama *et al.* (1968) showed that dehydrojuvabione, which is less active then juvabione, is increased in activity tenfold when the alicyclic rings are aromatized, and if the aliphatic double bonds are hydrochlorinated, the activity is again increased tenfold.

Thus, observations that metamorphosis is inhibited, or that embryonic development can be blocked by such juvenile hormonelike compounds, as well as similar findings, as previously discussed, for the ecdysones, are encouraging an intense search for natural products which may have been, for centuries, basic to certain insect–plant interactions.

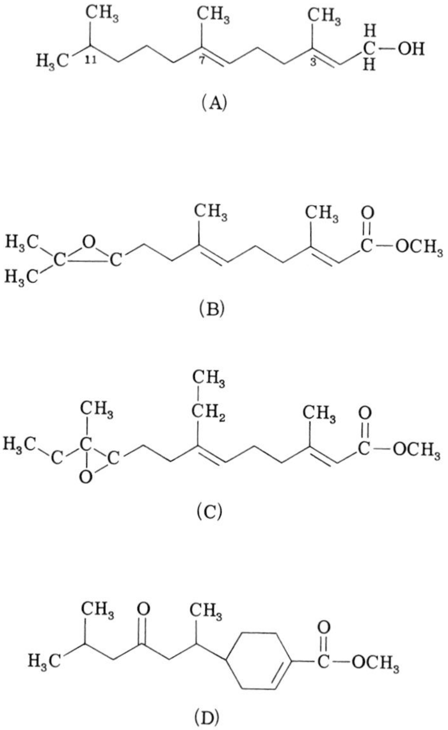

(A)

(B)

(C)

(D)

Fig. 11-13. Chemical structure of three compounds which possess juvenile hormone and gonadotropin activity. A. Farnesol, the basic compound with three isoprene units (15 carbons); B. 10,11-epoxyfarnesenic acid methyl ester (Bowers *et al.,* 1965); C. structural identity of the juvenile hormone of male *Hyalophora cecropia* (Röller *et al.,* 1967); D. Structural identity of juvabione, the active component from balsam firwood. (Bowers *et al.,* 1966.)

C. Physiology

It is apparent that the area of growth and development has attracted the largest share of interest in insect endocrinology. Wigglesworth's initial studies in this area, utilizing the exceptionally well-suited reduviid bug *Rhodnius prolixus,* has served as an excellent catalyst for many investigators who have subsequently produced a wealth of information on growth and development as well as other physiological systems (Beament and Treherne, 1967). Following an initial discussion of growth and development will be sections on reproduction, diapause, metabolism, and color adaptation. This group of five topics does not cover all aspects of insect endocrinology, for almost every insect organ system has been reported as being responsive in some way to a hormone. The Appendix should be consulted for additional endocrine relationships.

A number of reviews are available which deal specifically or in part with these five topics, as well as those omitted from our discussion. Two monographs on metamorphosis and diapause (Wigglesworth, 1959; Lees, 1955) are instructive, as well as the two recent books devoted exclusively to insect hormones (P. Joly, 1968; Novak, 1966).

1. GROWTH AND DEVELOPMENT

Insects, like all other arthropods, grow toward maturity through stages or instars. At the end of each instar, molting occurs when the old binding exoskeleton is shed, with the formation of a new skeleton of greater dimensions, thereby providing room for further body growth. On reaching the adult form, molting ceases, except in a few primitive insects from the orders Protura and Collembola. The final molt then becomes important, for it is the period of the insect's life cycle in which the adult characters are formed with the concurrent loss of specific juvenile structures. Metamorphosis is the term which refers to this period of change from juvenile to adult. If the juvenile has not deviated greatly from the adult characteristics, and if functional wings, mouth parts, and genitalia can be formed at the one final molt, then the insect illustrates hemimetabolous development or gradual metamorphosis as found in cockroaches, grasshoppers, and termites. On the other hand, when quite divergent juvenile structures appear, two molts, separated by an externally quiet pupal period, will be required for the formation of the adult, illustrating holometabolous development or complete metamorphosis, as in caterpillars, maggots, and grubs.

Of the two phenomena, molting and metamorphosis, which are basic to the growth and development of insects, molting can be discussed independently; but metamorphosis, since it involves exoskeleton renewal, must include some awareness of the molting process. In the following discussion the endocrinology of two phenomena will be treated as a unified series of events and the similarities rather than the differences between the Holometabola and Hemimetabola will be stressed (Fig. 11-14).

Three endocrine sources are instrumental in orchestrating the growth and development of insects: the brain, ecdysial glands, and corpora allata. The role of the neurosecretory cells in the brain as prime movers in both molting and metamorphosis has been made

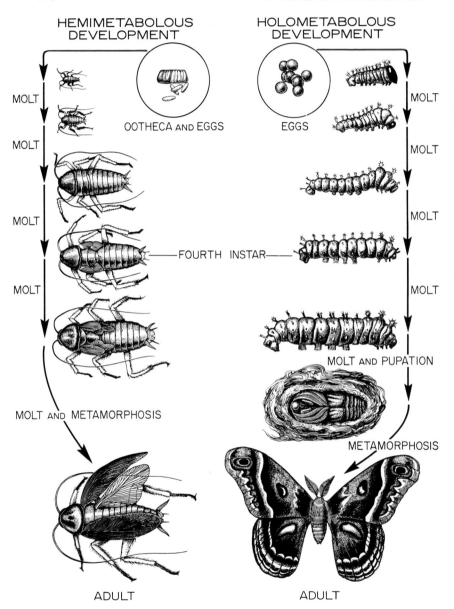

Fig. 11-14. Growth and development in the two contrasting types of life cycles. Hemimetabolous development, or incomplete metamorphosis, is illustrated by the cockroach and holometabolous development, or complete metamorphosis, by the silkworm. (From Turner, 1966.)

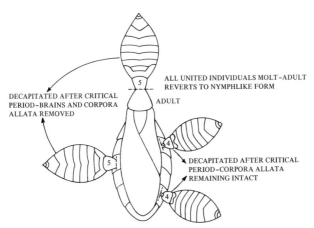

Fig. 11-15. The induction of a molt in an adult *Rhodnius* by utilizing the decapitation techniques of Wigglesworth. The broken lines indicate the level of decapitation; the blackened dot in the proximal end of the head represents the corpus allatum, and the number of the nymphal instar is indicated on the thorax. The decapitated adult is coupled with two fifth-stage nymphs, totally decapitated after the critical period, and with two fourth-stage nymphs partially decapitated after the critical period. The fifth nymphs provide the adult with ecdysone, whereas the fourth nymphs, having their corpora allata, provide juvenile hormone in addition to ecdysone. All the combined individuals molt; the fifth nymphs become supernumerary nymphs (sixth-stage), the fourth nymphs become fifth instars, and the adult reverts to a nymphlike form. (From Turner, 1966.)

abundantly clear. This research began with Kopec and his pioneer studies on the regulation of metamorphosis in caterpillars. His findings have been expanded into our present concepts by many researchers; but the two most influential are Wigglesworth (1964), with his extensive utilization of decapitation and parabiotic procedures on the hemimetabola *Rhodnius prolixus* (Fig. 11-15), and Williams, who utilized ablation and replacement techniques on the holometabola *Hyalophora cecropia* (Fig. 11-16).

Neurosecretory cells in the pars intercerebralis, upon proper stimulation, produce ecdysiotropin which passes to the neurohemal organ by axon transport. Evidence supporting this role for the brain has been extensive, with two major observations: (1) that the removal of the neurosecretory cells or their product by extirpation or ligation will interfere with molting; and (2) that upon the restoration of the essential brain factor by implantation or parabiosis normal function will be reinstated.

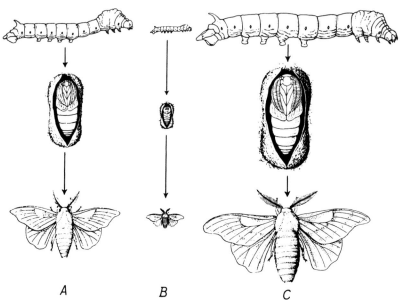

Fig. 11-16. Effects of the corpora allata on postembryonic development of *Bombyx mori*. A. Normal fifth instar, pupa and adult; B. allatectomized third instar, diminutive pupa, and diminutive adult; C. supernumerary larva produced from fifth instar by transplanting corpora allata of young larvae, giant pupa, and giant adult. (From Turner, 1966.)

The initial stimulus for the release of the ecdysiotropin in *Rhodnius* originates in the proprioceptors located in the abdomen. Upon the consumption of a large blood meal the stretch receptors are stimulated and the nerve impulses are sent to the brain through the ventral nerve cord. Presumably the neurosecretory cells are then stimulated to release the hormone. A similar pattern for brain stimulation, originating in proprioceptive organs located elsewhere on the body, has been suggested for several insects.

Upon its release from the neurohemal organ the neurosecretion ecdysiotropin stimulates the ecdysial glands to synthesize and secrete the very important molt-inducing hormone, ecdysone. Under the influence of this hormone the molting process of the insect is initiated.

It may be recalled that in crustaceans the ecdysial gland (y-organ) was inhibited in releasing the molting hormone, crustecdysone, by an eyestalk neurosecretion, while upon the ablation of both eyestalks, molting was initiated (Fig. 10-16). Thus, the same links are found in the chain controlling molting in both arthropod groups, but the neurosecretion plays opposite roles, i.e., inhibition in crustaceans and stimulation in insects.

The activation of the ecdysial gland prior to larval and pupal molts is generally believed to be via the ecdysiotropin released into the hemolymph from the cardiaca. This is undoubtedly the most common procedure in insects; however, three additional control systems have been suggested as possibilities: (1) innervation of the gland by one or several types of neurosecretory granules which originate in the ventral nerve cord; (2) stimulation by ecdysone itself; and (3) stimulation or inhibition by the juvenile hormone (Herman, 1968). How these factors stimulate the ecdysial gland cells to synthesize and release ecdysone is still unknown. However, an hypothesis concerning the mechanism of action of ecdysone will be mentioned later.

With the release of ecdysone into the hemolymph, the epidermis is stimulated to molt and, at the same time, the body tissues are caused to differentiate in the direction of adult structures. If, however, a third hormone, the juvenile hormone from the corpora allata, is released into the hemolymph, then the tissues will not become adult but will remain juvenile in form. This is the normal course of events during the insect's growth in size from the egg to the preadult form. This indicates that each insect cell is dependent upon multiple sets of genes which can be expressed successively during the life cycle with the production of immature or mature characters. The stimulus determining which set of genes will be expressed is the internal hormonal milieu, a varying concentration of juvenile hormone and the required amount of ecdysone for each molt. Consequently, when a high titer of juvenile hormone is present prior to each molt, the resulting organism will be immature; if the juvenile hormone is at a minimum, or completely absent, an adult will be formed and, with an intermediate level, the pupal period is attained (Fig. 11-17). Therefore, from an endocrinological point of view, the difference between molting and metamorphosis is the concentration of juvenile hormone in the hemolymph prior to each molt in the immature stages. The molt which occurs without any juvenile hormonal influence brings forth metamorphosis and the formation of the adult.

We mentioned that the evidence is substantial indicating that the ecdysial glands prior to each molt are stimulated by ecdysiotropin, but almost nothing is known about the control of the corpora allata at similar times in the immature insects. This gland may be under neuroendocrine control just as the ecdysial gland, or it may be more responsive to normal nervous stimulation. The former possibility certainly seems logical since axons with neurosecretory granules

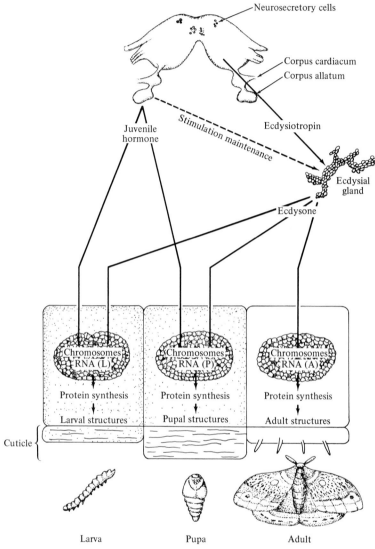

Larva   Pupa   Adult

Fig. 11-17. Endocrine influence at different periods in the life of a silkworm over its growth and development. It is assumed that a critical titer of ecdysone, and a specific concentration of juvenile hormone lead to the transfer of appropriate nuclear information to the cytoplasm via messenger compounds indicated as RNA "L," "P," or "A" for the formation of larval, pupal, or adult structures, respectively. The corpora allata may stimulate and maintain the ecdysial glands as well as regulate the synthesis of larval structures. (After Gilbert, 1964.)

have been observed in practically all allata examined ultrastructurally. The corpora allata of *Locusta migratoria* larvae have been reported to be stimulated by the lateral neurosecretory cells and inhibited by the medial cells of the protocerebrum (Girardie, 1965).

To escape the fate of continuing in the immature condition, the insect must be able to turn off or reduce the production of juvenile hormone prior to the last larval molt. This mechanism may be one of those mentioned above, and the inhibitory mechanism is a likely candidate, for, as we will see in the next section, the corpora allata becomes active again in the adult and exercises considerable control over reproductive development.

The question of how hormones influence the target cells and, thereby, the machinery that controls metabolic sequences has attracted much attention recently. Three possibilities have been proposed for the mode of action of hormones. The first is the hormone–enzyme interaction, the direct effect upon intracellular enzyme systems by the hormone. The second concerns the permeability of membranes, either at the cell surface or elsewhere within the cell, which may be altered by a hormone. The third, and most recent, is the hypothesis that the hormone may influence activation or suppression of genes on specific chromosomal loci. This last hypothesis has been developed largely from the observed relationship between ecdysone and the apparent activation of the giant chromosomes in the salivary gland cells of the midge *Chironomus*. This hypothesis has been based upon several observations which, together, interpret the appearance of "puffs" on chromosomes with DNA-dependent synthesis of messenger RNA. These areas of assumed activity, where the chromosome is reversibly loosened, unfolded, and puffed up, are normally observed in the midge prepupa at the time of tissue differentiation, which is also the time of a high ecdysone titer. When this relationship between chromosomal activation and increased hormone concentration is tested experimentally, the findings are positive. Within an hour following an injection of 10^{-10} gm of ecdysone into an intact last instar larva of *Chironomus*, two particular loci experience puffing. This is the earliest effect that has been found thus far for the injected molting hormone. Sequentially to these initial responses, other puffs occur; a significant rise in the messenger RNA also occurs. These nucleic acids direct protein biosynthesis with the formation of specific enzymes that serve specific metabolic pathways (Fig. 11-18) (Karlson, 1963; Karlson, 1965a,b).

Thus, even with the continued accumulation of evidence supporting the hormone–gene hypothesis, the question still exists, although at a

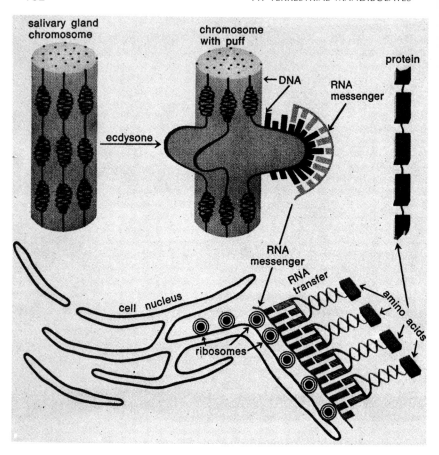

Fig. 11-18. Mechanism of action of ecdysone upon the giant salivary gland chromosomes of certain insects. The hormone is believed to act first by unfolding or loosening the DNA to produce a puff which is the site of RNA synthesis. The messenger RNA carries the information necessary for the alignment of amino acids in the synthesis of specific proteins (e.g., certain enzymes). These enzymes presumably evoke within the target cells the biochemical changes involved in the molting process. (From Turner, 1966, after Karlson, 1965b.)

finer level, concerning the nature of activation. Is there a direct combination of hormone and gene, or does the hormone influence the loci indirectly by way of removing a repressor or some other subcellular constituent from the chromosome? The possibility has also been suggested that the hormone may act by modifying the electrolyte composition of the nuclear sap. A recent review by Kroeger (1968) should be consulted for more detailed information on this last subject.

Thus, growth and development are controlled by three known hormones operating as a dual second-order neuroendocrine control mechanism: ecdysiotropin produced by neurosecretory cells in the brain and released to the hemolymph from the corpora cardiaca; ecdysone, produced in and released from the ecdysial glands in response to ecdysiotropin; and the juvenile hormone synthesized in and released from the corpora allata. The molting process, where a larger, immature insect is produced from a smaller nymph or larva, is under the control of both ecdysone which stimulates exoskeleton renewal, and juvenile hormone which favors the expression of immature characteristics. When the insect has reached a certain stage, ecdysone is released with either a reduced amount or no juvenile hormone. In response to this internal milieu the insect will metamorphose into either a pupa or an adult. The pupa of holometabolous insects will undergo an additional molt under the influence of ecdysone alone.

A hormone which is responsible for the tanning and hardening of the exoskeleton of newly molted insects has recently been reported from blowflies and roaches. In flies, the afferent stimulus originates probably in the abdomen within minutes of emergence and travels to the brain by way of the ventral nerve cord. There it triggers the release to the hemolymph of the protein neurohormone bursicon (Frankel and Hsiao, 1965). In the roach, the release of the tanning hormone is also under nervous control, but the site of release is the terminal abdominal ganglion, not the brain (Mills et al., 1965).

With the formation of the adult insect, the organism then becomes committed to species perpetuation. Hence, the endocrine control of reproduction will be discussed next.

2. REPRODUCTION

Some insects have evolved so that the adult, on emergence, immediately mates and oviposits. Their mouth parts are nonfunctional, so all energy required in the few hours of adult life is carried over from the larva. The reproductive system has obviously developed morphologically and functionally during the previous stage so that the abdomen, through the distention of the lateral and common oviduct, is filled with mature eggs. Mating and ovipositing are all that is expected of these adults for the continuation of the species. The requirements that such an adult might have for endocrine integration would concern the proper sequence and timing of these two events. On the other hand, the majority of adult insects reproduce in a cyclical manner during a life span which may extend for only a week; if a period of diapause intervenes, the adult may still be reproductive a year after emergence. Whether the period of reproduction extends only for a week or several months, there must be some hormonal control available so that the emerging adult, male or female, will feed, mature its gametes, mate successfully, and, in the case of the female, oviposit in a pattern that will provide some degree of success for the following generation.

The integrated control of reproduction in insects therefore involves much the same sequence of events as we have seen elsewhere; stimuli are directed into the central nervous system where neurosecretory cells are influenced to produce a change in the hormonal output of an endocrine organ. A very generalized statement such as this must be made with care, for there are so many fragments of knowledge, from the many insects which have been viewed in part and the few insects that have been studied in detail, concerning this area of insect endocrinology that exceptions are available which would invalidate each step in such a statement. Engelmann expresses this concern well in the concluding sentence of his recent (1968) review on the endocrinology of insect reproduction: "Any generalization on the principles involved is doomed to be revised within a few years and may not apply in the literal sense for any particular species."

Reflecting the incompleteness of our knowledge concerning insect reproduction and shying away from presenting fragments of detailed information, we will discuss, in this section, after briefly reviewing the morphology of the female reproductive system, the three aspects of neuroendocrine control of reproduction: stimuli which influence the brain, the role of the brain in controlling the corpora allata, and the mode of action of gonadotropin.

The basic functional units involved in egg development are the ovarioles, which consist of the terminal filament, germarium, egg tube, and pedicle. Ovarioles are in two groups, each bound by an epithelial sheath and thus forming a pair of ovaries. The ovariole pedicles of each ovary unite into the paired lateral oviduct, then meeting along the midline of the abdomen and forming the common oviduct. Through the gonopore, the common oviduct joins with the genital chamber, or copulatory bursa, which receives the male organ during copulation. The sperm, enclosed within a spermatophore or sperm packet, are retained within the seminal receptacle until required for fertilization. A pair of accessory glands serve principally to produce adhesive material needed for egg attachment.

The apex of the germarium houses the germ cells which undergo mitotic division forming three types of cells, the oocytes, tropic, and follicle. The oocytes, surrounded by follicular epithelium, move from

the germarium into the egg tube where vitellogenesis, or yolk formation, occurs, with a subsequent increase in egg size. When mature,
the oocytes pass out of the egg tube through the lateral, and into the
common oviduct. Fertilization occurs at the time of oviposition as the
egg passes through the genital chamber.

The nature of the stimuli which influence the brain, and thereby the
corpora allata, may be classified according to where they originate as
either extrinsic or intrinsic stimuli (Fig. 11-19). It is doubtful that any

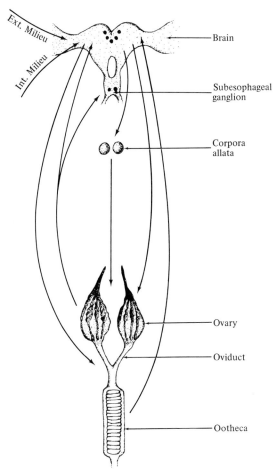

Fig. 11-19. Neuroendocrine integration by the brain of external and internal stimuli
in the control of reproductive processes of the adult female insect. (From de Wilde,
1964, after Scharrer, 1958.)

one factor acts independently on the neural integration; more likely several afferent stimuli are involved. The environmental temperature, photoperiod, humidity, and food supply have been implicated in a number of insects as major stimuli controlling adult reproduction. Since these stimuli have been most extensively studied in reference to diapause induction, further consideration of their influence upon the neuroendocrine system will be reserved for the next section on diapause.

Many intrinsic stimuli can be traced back to an extrinsic source such as nutritional factors in the hemolymph, reflecting the quality and quantity of the external food source. As in mosquitoes, it is the mechanical stimulus from a distended midgut following a large blood meal that initiates the chain reactions responsible for ovarian growth. Unless protein is provided in the diet of the blowfly *Calliphora erythrocephala* no mature ova are produced. The suggested series of events which are involved in this fly are most interesting. When the adult is given a choice of diet, it will choose one rich in protein rather than sugar. The dietary protein then stimulates the neurosecretory cells to bring about an increase in the protease of the gut. Thus, with the ability to digest proteins, metabolites appear in the hemolymph which directly stimulate the corpora allata to release its gonadotropin (Fig. 11–20). The stimulation of mating has been alluded to in studies on several insects, and it appears that the act of copulation is an effective signal which, when passed to the brain via the ventral nerve cord, is sufficient to bring about corpora allata activation. Whether this activation removes a neural inhibition of the glands or releases a stimulatory neurohormone has to be decided for each insect under investigation (Fig. 11-21).

Stimuli from the external and internal environment, impinging upon the brain, influence gonadal development by way of hormonal and nervous connections with the corpora allata. The most conclusive evidence that the neurosecretory cells are required for the activation of the corpora allata was provided by Thomsen (1952), who worked with *Calliphora erythrocephala,* and Highnam (1962), with *Schistocerca gregaria.* When the medial group of neurosecretory cells from these insects was removed, either surgically or by cautery, the corpora allata were rendered totally inactive. Following the implantation of active medial cells, the glands returned to an active state. A modification of this hormonal activation was reported by Johansson (1958) from his work on the milkweed bug *Oncopeltus fasciatus.* As partial development of the corpora allata and eggs was obtained in the

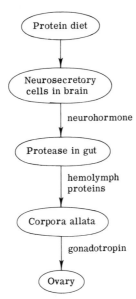

Fig. 11-20. Sequence followed in *Calliphora* for regulating protein metabolism. Following the intake of protein, the brain releases a neurohormone which causes the release of protease in the gut. The hemolymph protein increases and the corpora allata are activated to release gonadotropin which then stimulates vitellogenesis.

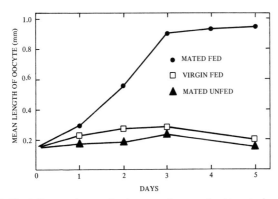

Fig. 11-21. The influence that mating has over the growth of terminal oocytes in the bedbug *Cimex* that had been fed and starved. (From Engelmann, 1968, after Davis, 1964.)

absence of the medial neurosecretory cells, something less than total dependence upon the neurosecretory cells was found; however, full growth of the glands and normal egg production were realized when the protocerebral cells were intact and undisturbed. Similar evidence as the above for *Calliphora, Schistocerca,* and *Oncopeltus* is now available on a number of other insects, indicating that the enlargement and activation of the corpora allata in some insects is dependent upon an "allatotropic" factor which very probably has its orgin in the medial cells of the protocerebrum.

Evidence for nervous control of the corpora allata, rather than hormonal, comes from work on the two ovoviviporous cockroaches *Leucophaea maderae* and *Diploptera punctata* (Engelmann, 1960). By severance in the adults of the nervi corporis cardiaci I, with or without the medial neurosecretory cells intact, nervous inhibition of the corpora allata is removed. When the same operation is accomplished in pregnant females with the brood sac still attached, the glands begin to increase in volume and egg maturation follows. With very selective brain damage, the regions which house the responsible cell bodies appear to be ventral to the medial neurosecretory cell region. Thus the corpora allata in a virgin, or otherwise quiescent female, would be under the influence of inhibitory stimuli originating in the brain. Upon the receipt of an appropriate stimulus, such as copulation or a particular mating behavior, the nervous inhibition would be removed and the development of the corpora allata would follow, with the subsequent release of gonadotropin. Starvation in female *Oncopeltus* causes inactivation of the corpora allata, producing a drop in the production of mature eggs. If the corpora allata are then severed from all brain connections, or if active corpora allata are implanted into starved females, ovarian maturation will be instituted.

From these observations it appears that if the corpora allata is directly controlled by the brain it may be either by nervous inhibition or hormonal stimulation. A combination of the two in which one can be overridden by the other, following a sufficiently strong stimulus, is also quite possible.

V. B. Wigglesworth was one of the first to link the endocrine system to the control of insect reproductive development. He showed, through careful experimentation, that the normal oocyte growth in *Rhodnius prolixus* is dependent upon a hormone produced by the corpora allata following adult emergence. This gland is the same structure that produces juvenile hormone in the immature insect; and, in fact, the two secretions, gonadotropin and juvenile hormone, are chemically identical. Wigglesworth's observations have been repeated in many insects, and now the role of the corpora allata in reproductive development of the female insect is relatively well established. It is the one link in the chain of events integrating various stimuli with reproductive responses which seem to be the most uniform in all insects. However, it appears not to act alone but, rather, to act in conjunction with a neurosecretory factor released to the blood from the corpora cardiaca. This neurohormone influences principally the metabolism of the fat tissues in such a way that the appropriate metabolites (proteins, carbohydrates, and lipids) are made available for the developing oocytes. It is at this point that gonadotropin becomes important, for it is believed that this hormone in some way facilitates the transfer of raw material from the hemolymph into the follicular cells, and then into the oocyte. When gonadotropin is not available, the oocytes are reabsorbed; when it is available, oocytes grow to maturity.

Gonadotropin has also been implicated in influencing the accessory glands of some male and female insects. In the female cockroach these glands are needed for the secretion of a material required in the construction of the ootheca.

In an insect which follows a cyclic reproductive pattern, gonadotropin must be inhibited between development of egg masses by various negative feedback mechanisms. One such mechanism was suggested by Nayar (1958) for the Pyrrhocoridae bug in which an "ovarian hormone" was believed to be secreted by the mature ovary, inhibiting further gonadotropin release and, at the same time, favoring the release of neurosecretion from the corpora cardiaca, thus promoting oviposition.

Several hypothetical schemes have been drawn to show the possible sequences of events appropriate to the control of reproduction in various insects (Fig. 11-22). The following outline was proposed by Hagadorn (1967) which appears to be the most general in its coverage.

1. In response to the appropriate stimuli, the median neurosecretory cells are activated and the inhibitory innervation to the corpora allata is suppressed. As a result, the corpora allata are induced to release their hormone. Protein metabolism is also stimulated.

2. Gonadotropin from the corpora allata promotes yolk formation in the ova and stimulates the sex accessory glands.

3. "Ovarian hormone" is released from the ovary containing maturing eggs. This inhibits further gonadotropin release from the corpora allata and favors the release of neurosecretory material from the corpora cardiaca.

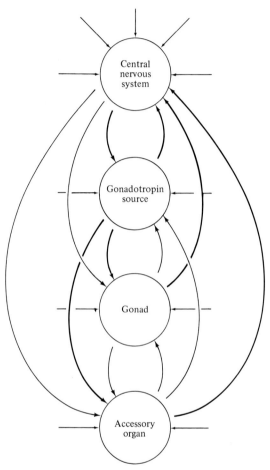

Fig. 11-22. Illustration of the possible routes for the integration of neural and hormonal mechanisms in the control of reproductive processes in insects. The heavy arrows represent major pathways. (From Scharrer and Scharrer, 1963.)

4. The neurosecretory materials promote oviposition, possibly by acting upon the central nervous system to remove neural inhibition of the centers controlling ovipositional movements.

5. Oviposition terminates the cycle, leaving the animal free to initiate the sequence again.

3. DIAPAUSE

The occurrence of an extended interruption in the normal progress of growth and development in insects is known as diapause. This term was first applied to a resting stage in grasshopper embryogenesis but later was broadened to refer to a condition of arrested growth in either the immature (embryo, larva, pupa) or the adult insect. Diapause now is best defined as a condition of reduced metabolism dependent upon and reflected in a number of physiological specializations; in this way it is set apart from quiescence, the more temporary or transitory reduction in metabolism, which is in response to the action of adverse environmental factors. The physiological specializations or characteristics associated with this phenomenon include a decrease in oxidative metabolism, body water, and responsiveness to external stimuli, and an increase in the insect's ability to withstand environmental stresses with hypertrophy of the fat tissue. Modification in the neuroendocrine system is believed to be paramount in all diapause studied thus far. Since diapause commonly serves as the overwintering condition for insects, the various stages are largely responsive to signals from the environment, such as photoperiod, temperature, humidity, or nutrition. These stimuli influence the neuroendocrine system, usually the neurosecretory cells in the protocerebrum, which in turn control the activity of various endocrine glands stimulating or inhibiting the organ systems.

Although a considerable portion of insects overwinter in the egg stage, only from *Bombyx mori* has substantial information been derived concerning the endocrinology of the egg or embryonic diapause. The silkworm is well suited for this research, not only because of its size but because in some races every generation of eggs enters a period of diapause, while in other races diapause occurs in response to well-defined environmental stimuli. When the developing embryos are placed under a long photoperiod and high temperature, the females produced from these embryos will deposit eggs which will enter a period of diapause. When a short photoperiod and low temperature are utilized for the embryos, the resulting female deposits a nondiapausing egg. The relationship between the summer environment and those conditions which produce overwintering eggs is evident.

Hasegawa (1957) was the first to show that the immediate stimulus for the development of diapause eggs is a neurosecretion, known now as the "diapause hormone," synthesized in and released from neurosecretory cells in the subesophageal ganglion of the female. This hormone is released so that it acts upon all eggs while still in the genital tract. Transplantation of subesophageal ganglia with active neurosecretory cells into pupae which are reared under the short photoperiod and low temperature conditions causes the mature females to lay diapause eggs. An extract, prepared from the activated brain–subesophageal ganglion complex, performed similarly when injected into pupae destined to produce nondiapausing eggs. The time for this injection is important. If it is within 2 days of adult emergence, the eggs will have already become determined; if injection occurs too early, the hormone is apparently inactivated by the pupal hemolymph. The influence of the brain over the subesophageal ganglion appears to be that of nervous inhibition which controls the release of the diapause hormone (Fukuda, 1962). This determination of diapause in the silkworm egg provides an excellent example of the delay which may occur between the receipt of a stimulus by the embryo and its demonstrable effect on the pupa. Nothing is known about where or how this information from the environment is retained by the neurosecretory cells in the embryonic brain and throughout larval development.

Larval or nymphal diapause has not received the intense study received by egg or pupal diapause, although possibly more individual species have been studied. From these reports several possible endocrine mechanisms have been proposed. One suggestion is that the corpora allata in *Chilo suppressalis* maintain some control over larval diapause by the continual release of an inhibitory substance. This is best illustrated by the termination of diapause in this larva following allatectomy (Mitsuhashi and Fukaya, 1960). In other larvae the hormonal control appears to be more similar to that found in pupal diapause where the neurosecretory cells of the brain and the ecdysial glands are instrumental in the initiation and termination of diapause. More support is needed for each of these hypotheses, but the second, which suggests a lack of molting hormone as the cause of diapause, appears to be the one which most easily fits with the current information on insect development.

It has been implied in the discussion thus far that the brain has been the transducer for the environmental stimuli influencing diapause induction. Although this is generally the arrangement, a possible exception occurs in the European corn borer larva *Ostrinia nubilalis*. This insect possesses a group of light- and temperature-sensitive secretory cells on the hind gut which, under the proper environmental conditions, are stimulated either directly or indirectly to release to the hemolymph the hormone proctodone. This secretion then activates the neurosecretory cells of the brain to release a neurohormone, presumably ecdysiotropin, which subsequently terminates diapause (Beck *et al.*, 1965). A number of important questions still exist concerning the stimulation of these endocrine cells as well as their relationship with the neuroendocrine system; nevertheless, a new pattern for the regulation of insect development has been uncovered (Fig. 11-23).

From the research efforts of C. M. Williams and many associates, there is today an extensive body of information concerning the physiology of pupal diapause derived almost entirely from the silkworms *Hyalophora cecropia* and *Antheraea pernyi* (Williams, 1956b; Williams *et al.*, 1965).

The period of dormancy in these insects extends from shortly after pupal formation in late summer for 6 or 8 months through the fall, winter, and into spring. As the temperature begins to rise, the insect completes its adult development, then emerges from the pupal cuticle and the cocoon. Diapause can be shortened considerably by placing the pupa under a low temperature environment for about six weeks. Experimentally, the same results can be obtained by the implantation of brains from activated cecropia into diapausing pupae or through parabiosis between an activated pupa and one or several in diapause.

From these observations, as well as many involving the ecdysial glands, it has been concluded that the inactive protocerebral neurosecretory cells of the diapausing pupae are stimulated to regain activity following a period of reduced temperature. The neurohormone

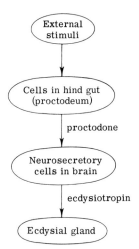

Fig. 11-23. The pattern followed by *Ostrinia nubilalis* in terminating larval diapause, which includes the secretion into the hemolymph of a hormone from the proctodeum.

released from these cells into the hemolymph stimulates the ecdysial glands to regain activity and release ecdysone, the molt-inducing hormone which terminates diapause. In fact, a simple injection of ecdysone into the pupa is sufficient to cause it to resume and complete adult development, initiate emergence, and bring about satisfactory mating and oviposition in the female. (Remember that the corpora allata is not necessary for reproduction in the adult silkworm.) Thus, basically the same endocrine pattern discussed earlier for molting is involved in the control of pupal diapause in Lepidoptera, but it is the method of stimulation and the response within the neurosecretory cells that differs. It had been suggested earlier that in the pupal brains of *Hyalophora* the absence of spontaneous electrical activity and the inability of the brain to respond to excitation were the basic causes for the dormancy of the neurosecretory cells and, therefore, were responsible for the start of the hormonal insufficiency which is expressed in the diapause condition. This theory has recently been subjected to serious criticism as both substantial cholinesterase and electrical activity have been recorded in *Hyalophora cecropia* as well as in several other lepidopteran species (Shappirio et al., 1967).

Many insects are able to enter a diapause condition as adults before the completion of reproductive development, but very little is known about the neuroendocrine system which regulates this phenomenon. There is, nevertheless, substantial evidence to suggest that photoperiod and temperature are the main environmental factors which control the function of the neurosecretory cells in an adult insect. Since the adults no longer retain their ecdysial glands, the deficiency of ecdysone, found to be the key endocrine in immature diapause, can hardly play the same role in the adult. The corpora allata, however, are present in the imago, and, in the few species studied thus far, the hormone deficiency appears to be the gonadotropin, required in the adult for development of mature ova. When active Colorado potato beetles *Leptinotarsa decemlineata* are allatectomized, they enter a condition which is quite similar to natural diapause. Reimplantation of the active glands restores the adult to its normal condition, but when this implantation of active glands is performed with naturally diapausing adults, there is no effect (de Wilde and Boer, 1961). Natural diapause in several adult insects has been broken by the application of the newly-synthesized juvenile hormones, which act here as gonadotropins.

The exact response of the neurosecretory cells to the environmental stimuli is not known; it is believed to be a failure of the synthetic mechanism in conjunction with some control of the release of prepared material. During diapause in some adults the neurosecretory cells contain a large amount of stainable material, while in others staining is difficult to observe. The significant observation is that in both conditions no observable transport of stainable material has been detected from the perikaryon to the axon terminals (Siew, 1965).

4. METABOLISM

Only concerning a few metabolic functions are there clear indications of neuroendocrine regulation, and much of this information is a product of studies concerning reproduction.

a. Carbohydrates and Lipids

Factors which regulate various aspects of carbohydrate metabolism appear to originate within certain neurosecretory cells of the protocerebrum and, through the connecting axons, are passed to the corpora cardiaca for storage. A hyperglycemic factor extractable from the cardiaca of several roaches increases the level of the disaccharide trehalose, the insect's blood sugar, in the hemolymph at the expense of glycogen in the fat tissue. This glycogenolytic effect is suspected to be a result of an increased activity of the enzyme phosphorylase within the fat tissue (Steele, 1963). A similar role for the medial neurosecretory cells of a mosquito has been suggested as exercising some degree of control over the synthesis of both glycogen and triglycerides (Van Handel and Lea, 1965).

Generally, lipid synthesis is increased following allatectomy, or under natural conditions when the corpora allata is inactivated to some degree, as in the preparatory stage of adult diapause. Two explanations have been provided for this; either there is a secretion favoring lipogenesis, or the buildup is a result of lack of lipid utilization by the reproductive system. This illustrates the correlative observations which have been made frequently, concerning many aspects of invertebrate endocrinology. For here a situation is observed reminiscent of the question posed in the second chapter concerning the histological interpretation of heavy staining within neurosecretory cells, where an increase in concentration can be explained as either an increased rate of production or a decreased rate of utilization.

b. Protein

Considerably more investigations have been conducted on protein metabolism in insects than on carbohydrates and lipids, but a clearer picture has not emerged. One reason for this is that two sources of hormones have been designated in different insects in the regulation of protein metabolism: neurosecretory cells in the brain, and the corpora allata. In *Colliphora* neurosecretion is vital for the release of protease into the gut. This causes an increase in the hemolymph protein content which in turn stimulates the corpora allata to release gonadotropin. Upon the removal of these neurosecretory cells, this ability to digest protein is clearly lost (Thomsen and Møller, 1963). A similar dependence upon neurohormones for synthesis of hemolymph proteins has been shown to exist for several other insects.

On the other hand, the brain does not seem to be important to *Rhodnius*, for when decapitation removes the brain but leaves the corpora allata, protein synthesis is not impaired (Coles, 1965). Similarly, in *Leucophaea* the synthesis of a specific hemolymph protein, believed to be involved in yolk formation, is not disturbed by brain cautery, but is lost following the removal or inactivation of the allata (Engelmann and Penny, 1966). This view has been strengthened further by studies on the cockroach *Nauphoeta* in which the corpora allata, not the neurosecretory cells and corpora cardiaca, have been shown to be responsible for the synthesis of hemolymph proteins (Lüscher, 1968).

c. Respiration

In addition to the influence of the corpora allata on the metabolism of several biochemical groups, there is evidence that gonadotropin also enhances respiration. This may be a very general response that can be correlated with morphogenetic change occurring throughout the body immediately following adult emergence, or a period of diapause.

It has also been suggested that the corpus allatum secretion specifically restores the coupling of respiration with phosphorylation at the termination of diapause (Stegwee, 1964). However, care must be exercised in preparing glandular extracts for injection, as pharmacologically active substances of many kinds may produce a significant change in short-term oxygen uptake.

d. Osmoregulation

Both diuretic and antidiuretic factors have been extracted from various parts of the nervous and neuroendocrine systems of different insects. Following a large blood meal, *Rhodnius* immediately begins to excrete via the Malpighian tubes much of the water which has passed from the gut into the body cavity, and within 2 to 3 hours the insect has thereby concentrated its intake to a fourth of its original

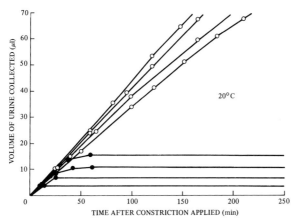

Fig. 11-24. The course of diuresis in freshly fed insects which were constricted either posterior to the neurosecretory cells (closed circles), thus retarding the flow of hormone into the abdomen, or anterior to the cells (open circles), thus allowing the hormone to reach the abdomen. (Maddrell, 1966.)

volume. This diuresis is in response to a diuretic hormone synthesized in secretory cells of the complex thoracic ganglion and released to the hemolymph from the surface of the abdominal nerves (Fig. 11-6). These cells are believed to be stimulated by stretch receptors in the abdominal wall which are activated following a large blood meal (Fig. 11-24) (Maddrell, 1966). An antidiuretic principle has been reported in *Periplaneta,* which is believed to be synthesized in the neurosecretory cells of the brain and stored in the corpora cardiaca. The ganglia of the ventral nerve cord also carry out similar activities. Nothing is known about the mode of action of either of these hormones on the Malpighian tubules. Berridge (1967), however, suggests that the diuretic hormone may regulate the entry of trehalose into the tubule cells, thus influencing the energy for active transport processes; this mechanism is not dissimilar to that suggested for insulin in mammalian cells.

5. COLOR ADAPTATION

Change in coloration may be considered a form of behavior which provides a protective coloration for the insect, most often the non-motile pupal stage (Highnam, 1964). There are two types of color adaptation, physiological or morphological. Physiological changes are rapid and short-term, and involve the movement of available pigment

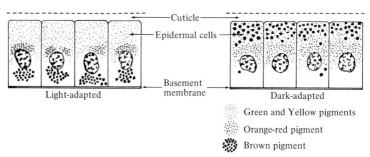

Fig. 11-25. Epidermal cells of the walking-stick *Dixippus morosus,* which illustrate physiological color change. The green and yellow pigments are not influenced by the environment and do not change their location in the cells, but the brown and orange-red pigments are concentrated in light-adapted insects and are dispensed nearer the cuticle in dark-adapted animals. (From Turner, 1966.)

to create a change in coloration. Chromatophores, which were common in crustaceans, have only been recorded in one insect; thus the migrating pigment is found within the normal epidermal cell. Morphological adaptations are slower and irrevocable, for they involve the formation of new pigment in the epidermal cells.

The phasmid *Carausius morsus* illustrates the first type quite well, and M. Raabe (1963) has provided much of the information on the hormonal controls of color change. This insect illustrates a diurnal pattern of color variation, light during the day and dark at night (Fig. 11-25). The darkening of the epidermal cells is stimulated by an active principle released to the hemolymph from the subesophageal ganglion, but having its source in the neurosecretory cells of the tritocerebrum. A second substance of weaker effect is obtained from the corpora cardiaca. The presence of such hormones can be illustrated easily by ligating the center of the body of a nymph *Carausius,* whereupon the anterior part of the body changes in color in the expected fashion, while the posterior portion remains constant in coloration. The presence in the brain of such stimulating factors has also been shown by brain transplants and the injection of brain extracts.

The environmental stimulus which influences the activity of neurosecretory cells is light, impinging upon the compound eyes. When the optic nerves are sectioned, the insect loses the rhythm of color change and becomes permanently darkened, which is similar to the response at night when no light is available. Thus, it is apparent that nervous impulses from the eyes travel into the brain and inhibit release of the darkening hormone. On the removal of the inhibition the hormone is released and the pigment migrates to the outer portion of the epidermal cells, providing a darker-colored exoskeleton. In insects a pigment-concentrating hormone has not been found.

Morphological color changes usually occur during ontogenetic development and the hormones which influence this color shift modify metabolic pathways involved in production of a number of different pigments.

In several papilionids the coloration of the pupa is dependent upon external conditions, such as the habitat of the last instar larva. The presence of dry twigs causes the brain to stimulate the ecdysial gland which releases a factor to the hemolymph that brings about darkening of the pupa by the deposition of melanin in the cuticle (Fig. 11-26). When the last instar larva is reared on green twigs, the ecdysial glands are not stimulated, no hormone is released, and, consequently, no melanin is formed. Thus the pupa remains a green color. This secretion from the ecdysial glands does not appear to be identical to the molt-inducing hormone ecdysone. On the contrary, other insects have linked their color adaptation, or pigment production, to the growth and development hormones. In this way, when a change occurs in the internal hormonal milieu by the normal alteration in ecdysone and juvenile hormone titers, or by experimentally altering the level of such hormones, a morphological color change is produced.

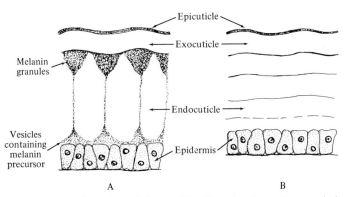

Fig. 11-26. Section of pupal exoskeleton of *Papilio xuthus* illustrating morphological color change. A. Melanin deposited in the outer region of the endocuticle thus giving a brown color; B. no melanin in the cuticle thus resulting in a transparent exoskeleton with the internal structures providing a green tint to the pupa. (From Highnam, 1964, after Hidaha, 1956.)

References

Beament, J. W. L., and Treherne, J. E., eds. (1967). "Insects and Physiology," Oliver & Boyd, Edinburgh and London.

Beck, S. D., Shane, J. L., and Colvin, I. B. (1965). Procotodone production in the European corn borer, *Ostrinia nubilalis*. *J. Insect Physiol.* **11,** 297–304.

Bern, H. A., and Hagadorn, I. R. (1965). Neurosecretion. *In* "Structure and Function in the Nervous System of Invertebrates" (T. H. Bullock and G. A. Horridge, eds.), Vol. 1, pp. 353–429. Freeman, San Francisco.

Berridge, M. J. (1967). Ion and water transport across epithelia. *In* "Insects and Physiology" (J. W. L. Beament and J. E. Treherne, eds.), pp. 327–347. Oliver & Boyd, Edinburgh and London.

Bowers, W. S. (1968). Juvenile hormone: Activity of natural and synthetic synergists. *Science* **161,** 895–897.

Bowers, W. S., Thompson, M. J., and Uebel, E. C. (1965). Juvenile and gonadotropic hormone activity of 10,11-epoxyfarnesenic acid methyl ester *Life Sci.* **4,** 2323–2331.

Bowers, W. S., Fales, H. M., Thompson, M. J., and Uebel, E. C. (1966). Juvenile hormone: Identification of an active compound from balsam fir. *Science* **154,** 1020–1021.

Butenandt, A., and Karlson, P. (1954). Uber die isoleirung eines metamorphosehormons der insekten in kristallisierter form. *Z. Naturforsch.* **96,** 389–391.

Cazal, P. (1948). Les glandes endocrines retrocérébrales des Insectes (étude morphologique). *Bull. biol. France Belgique,* suppl. **32,** 1–228.

Cerny, V., Dolejs, L., Labler, L., Sorm, F., and Slama, K. (1967). Dehydrojuvabione — a new compound with juvenile hormone activity from balsam fir. *Tetrahedron Letters* **12,** 1053–1057.

Coles, G. C. (1965). Studies on the hormonal control of metabolism in *Rhodnius prolixus* Stal. I. The adult female. *J. Insect Physiol.* **11,** 1325–1330.

Davis, N. T. (1964). Studies of the reproductive physiology of Cimicidae. I. Fecundation and egg maturation. *J. Insect Physiol.* **10,** 947–963.

de Wilde, J. (1964). Reproduction—endocrine control. *In* "The Physiology of Insecta" (M. Rockstein, ed.), Vol. 1, pp. 59–90. Academic Press, New York.

de Wilde, J., and Boer, J. A. (1961). Physiology of diapause in the adult Colorado beetle. II. Diapause as a case of pseudoallatectomy. *J. Insect Physiol.* **6,** 152–161.

Engelmann, F. (1960). Mechanisms controlling reproduction in two viviparous cochroaches. *Ann. N. Y. Acad. Sci.* **89,** 516–536.

Engelmann, F. (1968). Endocrine control of reproduction in insects. *Ann. Rev. Entomol.* **13,** 1–26.

Engelmann, F., and Penny, D. (1966). Studies on the endocrine control of metabolism in *Leucophaea maderae*. I. The hemolymph proteins during egg maturation. *Gen. Comp. Endocrinol.* **7,** 314–325.

Fraenkel, G., and Hsiao, C. (1965). Bursicon, a hormone which mediates tanning of the cuticle in the adult fly and other insects. *J. Insect Physiol.* **11,** 513–556.

Fraser, A. (1959). Neurosecretion in the brain of the larva of the sheep blowfly *Lucilia caesar. Quart. J. Microscop. Sci.* **100,** 377–394.

Fukuda, S. (1962). Hormonal control of diapause in the silkworm. *Gen. Comp. Endocrinol.* Suppl. 1, 337–340.

Gilbert, L. I. (1964). Physiology of growth and development: Endocrine aspects. *In*

"The Physiology of Insecta" (M. Rockstein, ed.), Vol. 1, pp. 149–255. Academic Press, New York.

Girardie, A. (1965). Contribution a l'étude du contrôle de l'activité des corpora allata par la pars intercerebralis chez Locusta migratoria (L.). Compt. Rend. Soc. Biol. 261, 4876–4878.

Gorbman, A., and Bern, H. A. (1962). "A Textbook of Comparative Endocrinology." Wiley, New York.

Hagadorn, I. R. (1967). Neuroendocrine mechanisms in Invertebrates. Neuroendocrinology (N. Y.) 1, 439–484.

Hasegawa, K. (1957). The diapause hormone of the silkworm Bombyx mori L. Nature 179, 1300–1301.

Herlant-Meewis, H., Naisse, J., and Nouton, J. (1967). Phénomènes neurosécretoires au niveau de la chaîne nerveuse chez les Invertébrés. In "Neurosecretion" (F. Stutinsky, ed.), pp. 203–218. Springer, Berlin.

Herman, W. S. (1967). The ecdysial glands of Arthropods. Intern. Rev. Cytol. 22, 269–347.

Herman, W. S. (1968). Control of hormone production in insects. In "Metamorphosis" (W. Etkin and L. I. Gilbert, eds.), pp. 107–141. Appleton, New York.

Highnam, K. C. (1962). Neurosecretory control of ovarian development in Schistocerca gregaria. Quart. J. Microscop. Sci. 103, 57–72.

Highnam, K. C. (1964). Hormones and behaviour in insects. Viewpoints Biol. 3, 219–255.

Ichikawa M., and Ishizaki H. (1961). Brain hormone of the silkworm, Bombyx mori. Nature 191, 933–934.

Ishizaki, H., and Ichikawa, M. (1967). Purification of the brain hormone of the silkworm Bombyx mori. Biol. Bull. 133, 355–368.

Jenkin, P. M. (1962). "Animal Hormones, A Comparative Survey, Part 1." Pergamon Press, Oxford.

Johansson, A. S. (1958). Hormonal regulation of reproduction in the milkweed bug, Oncopeltus fasciatus. Nature 181, 198–199.

Joly, P. (1968). "L'endocrinologie des Insectes." Masson, Paris.

Joly, R. (1962). Les glandes cérébrales, organes inhibiteurs de la mue chez les Myriapodes Chilopodes. Compt. Rend. 254, 1679–1681.

Karlson, P. (1963). New concepts on the mode of action of hormones. Perspectives Biol. Med. 6, 203–214.

Karlson, P. (1965a). "Mechanisms of Hormone Action." Academic Press, New York.

Karlson, P. (1965b). New concepts on the mode of action of hormones. Rassegna 42, 7.

Karlson, P., and Hoffmeister, H. (1963). Zur biogenese des ecdysons. I. Umwandlung von cholesterin in ecdyson. Z. Physiol. Chem. 331, 298–300.

Knowles, F. G. W. (1963). The structure of neurosecretory systems in invertebrates. Comp. Endocrinol. 2, 47–63.

Kobayashi M., and Kirimura, J. (1958). The brain hormone in the silkworm Bombyx mori L. Nature 181, 1217.

Kobayashi, M., and Yamazaki, M. (1966). The proteinic brain hormone in an insect, Bombyx mori L. Appl. Entomol. Zool. 1, 53–60.

Kobayashi, M., Kirimura, J., and Saito, M. (1962). Crystallization of the brain hormone of an insect. Nature 195, 515–516.

Kopec, S. (1917). Experiments on metamorphosis in insects. Bull. Acad. Sci. Cracovie pp. 57–60.

Kopec, S. (1922). Studies on the necessity of the brain for the inception of insect meta-
morphosis. *Biol. Bull.* **42,** 323–342.
Kroeger, H. (1968). Gene activities during insect metamorphosis and their control by
hormones. *In* "Metamorphosis" (W. Etkin and L. I. Gilbert, eds.), pp. 185–219.
Appleton, New York.
Lees, A. D. (1955). "The Physiology of Diapause in Arthropods." Cambridge Univ.
Press, London and New York.
Lüscher, M. (1968). Hormonal control of respiration and protein synthesis in the fat
body of the cockroach *Nauphoeta cinerea* during oocyte growth. *J. Insect Physiol.*
14, 499–511.
Maddrell, S. H. P. (1966). The site of release of the diuretic hormone in *Rhodnius*–a
new neurohemal system in insects. *J. Exptl. Biol.* **45,** 499–508.
Maddrell, S. H. P. (1967). Neurosecretion in insects. *In* "Insects and Physiology" (J. W.
L. Beament and J. E. Treherne, eds.), pp. 103–118. Oliver & Boyd, Edinburgh and
London.
Mills, R. R., Mathur, R. B., and Guerra, A. A. (1965). Studies on the hormonal control
of tanning in the American cockroach. I. Release of an activation factor from the
terminal abdominal ganglion. *J. Insect Physiol.* **11,** 1047–1053.
Mitsuhashi, J., and Fukaya, M. (1960). The hormonal control of larval diapause in the
rice stem borer, *Chilo suppressalis.* 3. Histological studies on the neurosecretory
cells of the brain and the secretory cells of the corpora allata during diapause and
post diapause. *Japan. J. Appl. Entomol. Zool.* **4,** 127–134.
Nayar, K. K. (1958). Studies on the neurosecretory system of *Iphita Limbata* Stal. V.
Probable endocrine basis of oviposition in the female insect. *Proc. Indian Acad.
Sci. (B),* **47,** 233–251.
Novak, V. J. A. (1966). "Insect Hormones." Methuen, London.
Palm, N. B. (1955). Neurosecretory cells and associated structures in *Lithobius forficatus*
L. *Ark. Zool.,* **9,** 115–129.
Prabhu, V. K. K. (1962). Neurosecretory system of *Jonespeltis splendidus* Verhoeff
(Myriapoda: Diplopoda). *In* "Neurosecretion" (H. Heller and R. B. Clark. eds.),
pp. 417–420. Academic Press, New York.
Raabe, M. (1963). Recherches experimentales sur la localistion intracerebrale du
facteur cromactif des Insectes. *Compt. Rend.* **257,** 1804–1806.
Robbins, W. E., Kaplanis, J. N., Thompson, M. J., Shortino, T. J., Cohen, C. F., and
Joyner, S. C. (1968). Ecdysone and analogs: Effects on development and repro-
duction of insects. *Science* **161,** 1158–1160.
Röller, H., Dohm, K. H. Sweely, C. C., and Trost, B. M. (1967). The structure of the
juvenile hormone. *Angew. Chem. Intern. Ed. Engl.* **6,** 176–180.
Scharrer, B. (1958). The role of neurosecretion in neuroendocrine integration. *In*
"Comparative Endocrinology" (A. Gorbman, ed.), pp. 134–147. Wiley, New York.
Scharrer, E., and Scharrer, B. (1963). "Neuroendocrinology." Columbia Univ. Press,
New York.
Shappirio, D. G., Eichenbaum, D. M., and Locke, B. R. (1967). Cholinesterase in the
brain of the cecropia silkmoth during metamorphosis and pupal diapause. *Biol.
Bull.* **132,** 108–125.
Siew, Y. C. (1965). The endocrine control of adult reproductive diapause in the chry-
somelid beetle *Galeruca tanaceti* (L.). II. *J. Insect. Physiol.* **11,** 463–479.
Slama, K., and Williams, C. M. (1965). Juvenile hormone activity for the bug *Pyrrhocoris
apterus. Proc. Natl. Acad. Sci. U.S.* **54,** 411–414.

Slama, K., Sucky, M., and Sorm, F. (1968). Natural and synthetic materials with insect hormone activity. 3. Juvenile hormone activity of derivatives of p-(1,5-dimethylhexyl) benzoic acid. *Biol. Bull.* **134,** 154–159.

Steele, J. E. (1963). The site of action of insect hyperglycemic hormone. *Gen. Comp. Endocrinol.* **3,** 46–52.

Stegwee, D. (1964). Respiratory chain metabolism in the Colorado potato beetle. II. Respiration and oxidative phosphorylation in "sarcosomes" from diapausing beetles. *J. Insect Physiol.* **10,** 97–102.

Thomsen, E. (1952). Functional significance of the neurosecretory brain cells and the corpus cardiacum in the female blowfly *Calliphora erthrocephala. J. Exptl. Biol.* **29,** 137–172.

Thomsen, E., and Møller, I. (1963). Influence of neurosecretory cells and of corpus allatum on intestinal protease activity in the adult *Calliphora erythrocephala* Meig. *J. Exptl. Biol.* **40,** 301–321.

Turner, C. D. (1966). "General Endocrinology." Saunders, Philadelphia.

Van Handel, E., and Lea, A. O. (1965). Medial neurosecretory cells as regulators of glycogen and triglyceride synthesis. *Science* **149,** 298–300.

Weber, H. (1949). "Grundriss der Insektenkunde," Jena. Gustav Fischer-Verlag.

Weyer, F. (1935). Uber drüsenartige Nervensellen im Gehirn der Honigbiene *Apis mellifera* L. *Zool. Anz.* **112,** 137–141.

Wigglesworth, V. B. (1959). "The Control of Growth and Form." Cornell Univ. Press, Ithaca, New York.

Wigglesworth, V. B. (1964). The hormonal regulation of growth and reproduction in insects. *Advan. Insect Physiol.* **2,** 247–336.

Williams, C. M. (1956a). The juvenile hormone of insects. *Nature* **178,** 212–213.

Williams, C. M. (1956b). Physiology of insect diapause. X. An endocrine mechanism for the influence of temperature on the diapausing pupa of the cecropia silkworm. *Biol. Bull.* **110,** 201–218.

Williams, C. M. (1967). The present status of the brain hormone. *In* "Insects and Physiology" (J. W. L. Beament and J. E. Treherne, eds.), pp. 133–139. Oliver & Boyd, Edinburgh and London.

Williams, C. M., and Robbins, W. E. (1968). Conference on insect-pest interactions. *Bioscience* **18,** 791–799.

Williams, C. M., Adkisson, P. L., and Walcott, C. (1965). Physiology of insect diapause. XV. The transmission of photoperiod signals to the brain of the oak silkworm, *Antheraea pernyi. Biol. Bull.* **128,** 497–507.

APPENDIX

TABLE A-1
Classification Used in Study[a]

Kingdom	ANIMAL	
Subkingdom	PROTOZOA	Protozoa
	METAZOA	
Branch A	MESOZOA	Mesozoa
Branch B	PARAZOA	Proifera
Branch C	EUMETAZOA	
Grade I	RADIATA	Cnidaria (Chap. 3)
		Ctenophora
Grade II	BILATERIA	
Division A	PROTOSTOMIA	
Subdiv. 1	ACOELOMATES	
Superphylum	ACOELOMATA	Platyhelminthes (Chap. 4)
		Turbellaria
		Trematoda
		Nemertinea (Chap. 5)
Subdiv. 2	PSEUDOCOELOMATES	Acanthocephala
		Entopocta
Superphylum	ASCHELMINTHES	Rotifera
		Gastrotricha
		Kinorhyncha
		Priapulida
		Nematoda (Chap. 6)
		Nematomorpha
Subdiv. 3	COELOMATES	
Superphylum	TENTACULATA	Phoronida
		Ectoproeta
		Brachiopoda

Superphylum	INARTICULATA	Sipuneuloidea
		Mollusca (Chap. 7)
		Gastropoda
		Pelecypoda
		Cephalopoda
Superphylum	ARTICULATA	Echiurpidea
		Annelida (Chap. 8)
		Polychaeta
		Oligochaeta
		Hirudinea
		Tardigrade
		Onychophora
		Pentastomida
		Arthropoda
Subphylum		Chelicerata (Chap. 9)
		Merostomata
		Arachnida
		Mandibulata (Chap. 10, 11)
		Crustacea
		Chilopoda
		Synphyla
		Diplopoda
		Insecta
Division B	DEUTEROSTOMIA	Echinodermata
		Chaetognatha
		Pogonophora
		Hemichordata
		Chordata

[a] After Meglitsch (1967).

SUPPLEMENTARY READINGS

Listed below are selected laboratory exercises in invertebrate endocrinology which have been published and are easily available.

A. "Experiments in Physiology and Biochemistry" (1968). G. A. Kerkut, ed. Academic Press, New York.
   1. Chen, D. H. Allatectomy of the American Cockroach *Periplaneta americana* (L.), 8 pp.
   2. Maddrell, S. H. P. Hormonal Control of Excretion in an Insect (*Rhodnius prolixus*), 11 pp.

B. Welsh, J. H. and Smith, R. I. (1960). "Laboratory Exercises in Invertebrate Physiology." Burgess Publishing Company, Minneapolis.
   1. Control of Color Change in Crustaceans, 2 pp.
   2. Control of Retinal Pigment Movement in Crustaceans, 1 p.
   3. Control of Molting in Crustaceans, 1 p.

C. Zarrow, M. X., Yochim, J. M., McCarthy, J. L., and Sanborn, R. C. (1964). "Experimental Endocrinology, a Sourcebook of Basic Techniques." Academic Press, New York.
   1. Puparium Formation of Diptera.
   2. Removal of the Lepidopteran Brain.
   3. Juvenile Hormone and Allatectomy of the Cockroach.
   4. Preparation of Isolated Abdomens of Lepidoptera.
   5. Crustacean Eyestalks and Molting.
   6. Eye pigment Migration in Crustacea.
   7. Control of Chromatophores in Crustacea.
   8. Endocrine Control of Rhythmic Activity in Blattaria.
   9. Control of Timing in the Rhythmic Activity of Blattaria.
   10. Role of the Subesophageal Ganglia in Rhythmic activity.
   11. Caste Determination in Termites.

AUTHOR INDEX

Numbers in italics refer to the pages on which the complete references are listed.

207

SUBJECT INDEX

A

Acarina, 108
Accessory glands, 189
Acetylcholine, 5
Acetylcholinesterase, 28
Acoelomate, 33, 43
Activation hormone, 169
Adrenaline, 5
Adult diapause, 173, 194, 195
Aestivitation, 93
Aestivo-hibernation, 95
Afferent pathways, 9
"Allatotropic" factor, 188
Amphipoda, 128
Amphineura, 54
Amphiporus pulcher, 44
Androgenic glands, 23, 118, 130, 144–147
Annelida, 73–104
Annual cycle (period), 38, 63, 95, 102, 106, 117, 155
Antheraea pernyi, 193
Antidiuretic hormone, 149, 196, 197
Apis mellifera, 156
Apterygote, 155, 159, 162, 165
Arachnida, 106, 108
Araneae, 109

Arion, 64
Arthropoda, 105, 115, 153
Ascaris lumbricoides, 48, 50
Aschelminthes, 47
Atoke, 78
Axis cylinders, 15
Axonal transport of neurosecretion, 18, 22
Azan, 5

B

Barrington, E. J. W., 131
Blood sugar
  levels, 148
  regulation, 149
Bombyx mori, 160, 169
Bothrioplana, 34
Brachyura, 126, 128
Brain hormone, 169
Branchiopoda, 116
Bursicon, 183

C

Calcium deposition, 148
Callinectes, 143
Calliphora erythrocephala, 186, 187, 196

211

216

Pulmonata, 59–64
Pupal diapause, 193, 194
Pyrrhocoris, 173

Q

Quiescence, 191

R

Reflecting pigment, 135–137
Regeneration, 28–32, 36–37, 75, 78, 85–
89, 96–97
"Regeneration hormone," 89
Release of neurosecretion, 19–21
Reproduction, 32, 38, 44, 57, 64–70, 75,
80–85, 92–93, 102, 109, 112, 144–
148, 183–190
Reproductive inactivity, 56–57, 93, 147,
194
Respiration, 139, 148, 196
Retinal pigment migration, 135–137
Rhinophores, 57
Rhodnius prolixus, 9, 13, 174, 177, 188,
196
Rhynchodaeum, 43
Ribbon worms, 43
Rostellum, 41

S

Samia cynthia ricini, 169
Scallops, 64
Scaphopoda, 54
Schistocerca gregaria, 186
Scolex, 41
Scorpionida, 108
Scorpions, 105
Second-order neuroendocrine reflex, 12,
58, 183
Secondary organ of Schneider, 109
Sedentaria, 75
Sensory pore x-organ, 118, 122, 124

Sex reversal, 146
Sexual differentiation, 144
Sexual maturation, 56, 83, 109
Shrimp, 117, 132
Sinus gland, 22, 57, 116, 118, 122, 124,
128, 133, 137, 142, 147
Snails, 55
Somatic pigmentation, 131–135
Somatic transformation, 75
Spawning, 80, 84–85
Spermatogenesis, 64, 81–82, 145
Spider crab, 138
Spiders, 105, 109
Squid, 66
Squilla mantis, 126–128
Starvation, 188
"Stomatogastric ganglion," 108
Strobilization, 41
Synapse, 5
Synaptic-like vesicles, 20
Synaptic vesicles, 15
Synthesis of neurosecretion, 17–18

T

Tanning hormone, 183
Tapeworm, 40
Technique, *in situ,* 16
Temperature, 191, 194
Tentacles, 31
Theromyzon, 99–102
Third-order neuroendocrine reflex, 13
Ticks, 105, 108
Trehalose, 195
Trematoda, 39–40
Trilobite, 105
Tryptophan, 139
Turbellaria, 33–39
Tubercula pubertatis, 93

V

Vitellogenesis, 81, 83, 85, 93, 147,
185